X-mas 1988
Daddy from Jan

D0687359

The Quiet Crisis
and the
Next Generation

The Quiet Crisis and the Next Generation

STEWART L. UDALL

GIBBS·SMITH
P
PUBLISHER

PEREGRINE SMITH BOOKS

SALT LAKE CITY

First edition

92 91 90 89 88   3 2 1

Published by Gibbs Smith, Publisher, P.O. Box 667, Layton, Utah 84041

Grateful acknowledgment is made to Henry Holt and Company, Inc., for selec-
tions from "The Gift Outright" from *The Poetry of Robert Frost*, edited by
Edward Connery Lathem, copyright 1942 by Robert Frost; "Wilderness" from
Cornhuskers by Carl Sandburg, copyright 1918 by Holt, Rinehart and Winston,
Inc., renewed 1946 by Carl Sandburg, reprinted by permission of Harcourt Brace
Jovanovich, Inc.; and "The Ballad of William Sycamore" from *Ballads and Poems*
by Stephen Vincent Benét, copyright 19831 by Stephen Vincent Benét, copy-
right 1931 by Stephen Vincent Benét, copyright renewed © 1959 by Rosemary
Carr Benét.

And to Houghton Mifflin Company, Boston, Massachusetts, for the use of
four lines from "Wildwest" in *New and Collected Poems 1917-1952* by Archibald
MacLeish. Copyright © 1985 by The Estate of Archibald MacLeish. Reprinted
by permission of Houghton Mifflin Company.

And to Macmillan Publishing Company, New York, New York, for the use
of thirteen lines from "The Ghosts of Time," (copyright 1935 by Charles
Scribner's Sons, renewal copyright © 1963 by Paul Gitlin) from *A Stone, A Leaf,
A Door* by Thomas Wolfe.

Cover design by J. Scott Knudsen

Printed and bound in the United States of America

Library of Congress Cataloging-in-Publication Data
Udall, Stewart L.
  The quiet crisis and the next generation.

  Updates: The quiet crisis. 1963.
  Includes index.
  1. Conservation of natural resources—United States—History. 2. Ecology—
United States—History. 3. Environmental protection—United States—History.
I. Title.
S930.U3   1988   333.7′2′0973   88-15123
ISBN 0-87905-333-X

For Lee

CONTENTS

FOREWORD

THE HISTORY of America is, more than that of most nations, the history of man confronted by nature. Our story has been peculiarly the story of man and the land, man and the forests, man and the plains, man and water, man and resources. It has been the story of a rich and varied natural heritage shaping American institutions and American values; and it has been equally the story of Americans seizing, using, squandering and, belatedly, protecting and developing that heritage. In telling this story and giving this central theme of American history its proper emphasis and dignity, Secretary Udall puts us all in his debt.

From the beginning, Americans had a lively awareness of the land and the wilderness. The Jeffersonian faith in the independent farmer laid the foundation for American democracy; and the ever-beckoning, ever-receding frontier left an indelible imprint on American society and the American character. And Americans pioneered in more than the usual way. We hear much about "land reform" today in other parts of the world; but we do not perhaps reflect enough on the extent to which land reform, from the Northwest Ordinance through the Homestead Act of the Farm Security Administration and beyond, was an American custom and an American innovation.

Yet, at the same time that Americans saluted the noble bounty of nature, they also abused and abandoned it. For the first century after independence, we regarded the natural environment as indestructible—and proceeded vigorously to destroy it. Not till the time of Marsh and Schurz and Powell did we begin to understand that our resources were not inexhaustible. Only in the twentieth century have we acted in a systematic way to defend and enrich our natural heritage.

The modern American record in conservation has been brilliant and distinguished. It has inspired comparable efforts all around the earth. But it came just in time in our own land. And, as Mr. Udall's vivid narrative makes clear, the race between education and erosion, between wisdom and waste, has not run its course. George Perkins Marsh pointed out a century ago that greed and shortsightedness were the natural enemies of a prudent resources policy. Each generation must deal anew with the "raiders," with the scramble to use public resources for private profit, and with the tendency to prefer short-run profits to long-run necessities. The nation's battle to preserve the common estate is far from won.

Mr. Udall understands this—and he understands too that new times give this battle new forms. I read with particular interest his chapter on "Conservation and the Future," in which he sets forth the implications for the conservation effort of the new science and technology. On the one hand, he notes, science has opened up great new sources of energy and great new means of control. On the other hand, new technical processes and devices litter the countryside with waste and refuse, contaminate water and air, imperil wildlife and man and endanger the balance of nature itself. Our economic standard of living rises, but our environmental standard of living—our access to nature and respect for it—deteriorates. A once beautiful nation, as Mr. Udall suggests, is in danger of turning into an "ugly America." And the long-run effect will be not only to degrade the quality of the national life but to weaken the foundations of national power.

The crisis may be quiet, but it is urgent. We must do in our own day what Theodore Roosevelt did sixty years ago and Franklin Roosevelt thirty years ago: we must expand the concept of conservation to meet the imperious problems of the new age. We must develop new instruments of foresight and protection and nurture in order to recover the relationship between man and nature and to make sure that the national estate we pass on to our multiplying descendants is green and flourishing.

I hope that all Americans understand the importance of this effort, because it cannot be won until each American makes the preservation of "the beauty and the bounty of the American earth" his personal commitment. To this effort, Secretary Udall has given courageous leadership, and, to this understanding, *The Quiet Crisis* makes a stirring and illuminating contribution.

JOHN F. KENNEDY

INTRODUCTION

History is not truth.
Truth is in the telling.
—ROBERT PENN WARREN
"Wind and Gibbon"

WHEN I wrote the foreword to *The Quiet Crisis* twenty-five years ago this summer, I used these words to describe my reaction to the predicament facing my generation:

> One week last fall two events came to my attention which seemed to sum up the plight of modern man: the first was a press report which indicated that T. S. Eliot, the poet, was a victim of London's latest "killer fog" and lay gravely ill; the second was a call from a preservation-minded citizen of New Hampshire who informed me that Robert Frost's old farm—fixed for all time in memory by the poem "West-running Brook"—was now an auto junk yard.
>
> The coincidence of these two events raised questions in my mind: Is a society a success if it creates conditions that impair its finest minds and make a wasteland of its finest landscapes? What does material abundance avail if we create an environment in which man's highest and most specifically human attributes cannot be fulfilled?

The Quiet Crisis was a call to action, an effort to combat apathy by exciting interest in our conservation heritage. I wanted to arouse concern about the spread of environmental degradation, to help the conservation movement focus on new goals, and to remind my country of the ideas and ideals that had inspired earlier generations of Americans to see themselves as stewards of resources that belonged to their children and to the unborn.

It is fascinating now to look back and survey the situation that existed in Washington in the spring of 1963. Expansive projects dominated the national scene in those days. The impending "conquest" of outer space—and the nuclear scientists' assurances that limitless inexpensive energy would soon propel humankind into an era of superabundant resources—aroused expectation that caused someone to describe that period as "the soaring sixties." Optimism about our technological prowess had a powerful grip on American thought during the Kennedy years. Indeed, that buoyancy explains some of the extravagant language I used in my chapter, "Conservation and the Future," to describe the bold plans and projects of the "new science." In those days, Americans of my generation were conditioned to believe that the same kinds of minds that had unlocked the secrets of the atom could achieve additional miracles that would transform life on earth.

At the time, the great men of atomic science still peered down from pedestals. Ordinary citizens had no reason to question their prophecies that by the 1980's breeder reactors and fusion power plants would provide "free" energy that would, in due course, eliminate the resource shortages that had long hindered human advancement. Yet today it is downright embarrassing for me to read some of the euphoric language I used to depict the "new world" these scientists were dangling before us.

A quarter of a century later, I am intrigued by the circumstance that, while the earth-changing plans and projects of nuclear scientists and engineers were unraveling, the values and concepts involving respect for nature's constraints advocated by ecologists were quietly altering our thinking about the future. I began to see in the 1970's that the rise of environmentalism and the faltering of Big Technology were part of a single, interacting episode of history. I also began viewing this whole matter as a competition in which ecology, a tenacious earthbound tortoise, outran a hare that called itself Big Science.

In the years after 1963 this competition ranged over many areas of our national life. There were conflicts about certain scientists' plans to "conquer" nature, about the local and global environmental impacts of industrial effluents, about the nuclear power option, about the limits to growth imposed by nature's ecosystems, about energy conservation as an appropriate response to petroleum shortages, about applications of biotechnology, and about specific projects such as the SST, a commercial supersonic airplane.

Because I had helped orchestrate the beginning of the environmental movement and had participated in many of the disputes that made it a major new political and social cause in this country, friends urged me to update *The Quiet Crisis* in order to share with a new generation the ecological insights that had transformed the conservation cause into a latter-day environmentalism that has become a dominant strain of American thought.

In seeking to accomplish this task, I made two decisions that influenced the format of this book. I decided to find a publisher who would, under one cover, reprint the full text of *The Quiet Crisis* along with my new text. And I also decided to follow the pattern of my earlier work by composing profiles of leaders who influenced the evolution of the ecological cause from 1963 to the present.

Plainly, Americans today need clarification about the origins of environmental thought. *The Quiet Crisis and the Next Generation* shows how a movement that spoke for the concerns of a minority of Americans in 1963 evolved into a cause that now commands broad national support on most environmental issues. I explain how the ideas and values of environmentalists took precedence over the schemes of Big Technology, that, not long ago, were perceived as a wave of the future. Among the stories I tell is the rise, in two swift decades, of a sophisticated system of environmental law in the United States. I am convinced that prudent economics in 1988 dictate that a national lifestyle formerly tied to energy waste must embrace the conservation ethic.

This update, then, is an effort to examine the people, ideas, and events that have generated a spacious concept of land stewardship in the past quarter of a century. And, since the central truths that govern life on earth do not change, the summons I offer to the oncoming generation is the same challenge I presented to my contemporaries in the words I wrote twenty-five years ago:

> Each generation has its own rendezvous with the land, for despite our fee titles and claims of ownership, we are all brief tenants on this planet. By choice, or by default, we will carve out a land legacy for our heirs. We can misuse the land and diminish the usefulness of resources, or we can create a world in which physical affluence and affluence of the spirit go hand in hand.

STEWART L. UDALL

Phoenix, Arizona
May 1988

PART ONE

THE QUIET CRISIS

CHAPTER I

The Land Wisdom
of the Indians

*In the dust where we have buried the silent races and their
abominations we have buried so much of the delicate magic
of life.*

—D. H. LAWRENCE (at Taos)

THERE ARE, today, a few wilderness reaches on the
North American continent—in Alaska, in Canada, and in
the high places of the Rocky Mountains—where the early-
morning mantle of primeval America can be seen in its
pristine glory, where one can gaze with wonder on the land
as it was when the Indians first came. Geologically and
geographically this continent was, and is, a masterpiece.
With its ideal latitude and rich resources, the two-billion-
acre expanse that became the United States was the promised land for active men.

The American continent was in a state of climax at the
time of the first Indian intrusions ten millennia or more ago.
Superlatives alone could describe the bewildering abundance of flora and fauna that enlivened its landscapes: the
towering redwoods, the giant saguaro cacti, the teeming
herds of buffalo, the beaver, and the grass were, of their
kind, unsurpassed.

The most common trait of all primitive peoples is a reverence for the life-giving earth, and the native American shared this elemental ethic: the land was alive to his loving touch, and he, its son, was brother to all creatures. His feelings were made visible in medicine bundles and dance rhythms for rain, and all of his religious rites and land attitudes savored the inseparable world of nature and God, the Master of Life. During the long Indian tenure the land remained undefiled save for scars no deeper than the scratches of cornfield clearings or the farming canals of the Hohokams on the Arizona desert.

There was skill in gardening along with this respect for the earth, and when Sir Walter Raleigh's colonists came warily ashore on the Atlantic Coast, Indians brought them gifts of melons and grapes. In Massachusetts, too, Indians not only schooled the Pilgrims in the culture of maize and squashes, but taught them how to fertilize the hills with alewives from the tidal creeks. The Five Nations and the Algonquians of the Northeast; the Creeks, Choctaws, Chickasaws, Cherokees, and Seminoles of the South; the village-dwelling Mandans of the Missouri River country; the Pueblos of Hopi, Zuni, and the Rio Grande; and the Pima of the Southwest, all put the earth to use and made it bring forth fruit. Their implements were Stone Age, but most tribes were acquiring the rudiments of a higher civilization. They were learning how to secure a surplus from the earth, and were beginning to invest it in goods, tools, and buildings, and to devote their leisure hours to craft and art work and to the creation of religious rites and political systems.

The idea has long been implanted in our thinking that all American Indians belonged to nomadic bands that developed neither title to, nor ties with, the land. This is misconceived history, for even the tribes that were not village dwellers, tending garden plots of corn, beans, or cotton, had stretches of land they regarded as their own. But there was a subtle qualification. The land and the Indians were bound together by the ties of kinship and nature, rather than by an understanding of property ownership. "The land is our mother," said Iroquois tradition, said the Midwest Sauk and Foxes, said the Northwest Nez Perces of Chief Joseph. The corn, fruits, roots, fish, and game were to all tribes the gifts which the Earth Mother gave freely to her children. And with that conception, the Indian's emotional attachment for his woods, valleys, and prairies were the very essence of life.

The depth of this feeling is reflected in the Navajos, who scorned the rich Oklahoma prairie country offered them by the government, and chose to live in their own arid and rugged deserts. It is reflected also in the Cherokees who, in the space of one generation, changed their whole way of life, established schools and libraries, produced an alphabet, planned a constitution and a legislature, and went to work in their own mills and blacksmith shops—all with the purpose of becoming so civilized that the whites would allow them to stay on their own lands and not ship them west to the Territories.

To the Indian, the homeland was the center of the universe. No member of a civilized people ever spoke of his native land with more pride than is apparent in the speech of the Crow Chief, Arapooish: "The Crow country," he said, "is exactly in the right place. It has snowy mountains and sunny plains; all kinds of climates and good things for every season. When the summer heat scorches the prairies, you can draw up under the mountains, where the air is sweet and cool, the grass fresh, and the bright streams come tumbling out of the snowbanks. There you can hunt the elk, the deer, and the antelope, when their skins are fit for dressing; there you will find plenty of white bears and mountain sheep.

"In the autumn, when your horses are fat and strong from the mountain pastures, you can go down into the plains and hunt the buffalo, or trap beaver in the streams. And when winter comes on, you can take shelter in the woody bottoms along the rivers; there you will find buffalo meat for yourselves, and cottonwood bark for your horses; or you may winter in the Wind River Valley where there is salt weed in abundance.

"The Crow country is exactly in the right place. Everything good is to be found there. There is no country like the Crow country."

Here is affection for the land, but no notion of private ownership. The idea that land could be bought and sold was an alien concept to the Indians of America. They clung possessively to certain chattels, but lands were nearly always held in common. An individual might have the use of a farm plot, but at his death it reverted back to the community.

The confrontation of Indians and whites had in it the seeds of hopeless misunderstanding from the start. The two cultures had produced irreconcilable concepts of landownership, and once the first white man set foot on American soil, the drama unfolded with all the certain sweep of a Greek tragedy.

Englishmen, especially, coveted land. It was something to be owned outright. Had not the English King given the charter deeds? The sixteenth-century Spaniard, by contrast, was not primarily interested in seizing land: the soldier wanted personal plunder; the priest came with his seeds and livestock to save Indian souls.

To the joint-stock companies of Virginia, intent on commercial profits, and to the colonizing Pilgrims, exclusive possession was the be-all and end-all of landownership. But the Indian's "title," based on the idea that he belonged to the land and was its son, was a charter to use—to use in common with his clan or fellow tribesmen, and not to *use up*. Neither white nor Indian fully grasped the concept of the other. The Indian wanted to live not just in the world, but with it; the white man, who thought in terms of estates and baronies, wanted land he alone could cultivate and use.

In the beginning, friendship and co-operation with the Indians were essential if the colonists were to gain a foothold in America, for the white man was badly outnumbered. To be unneighborly was to risk violence, and respect for Indian rights was the better part of wisdom. The upright conduct of the first colonists in Massachusetts and Virginia drew generous response from powerful chiefs who helped the settlements survive.

Live and let live was the inevitable opening keynote, for muskets could neither cut trees nor keep the peace. In the meeting of alien worlds both Indians and whites had something to learn from each other, and if the newcomers borrowed the idea of a feast of thanksgiving from a harvest celebration of neighboring Indians, so much the better.

But the first phase ended quickly, and as stockades were completed and new colonists swelled the ranks of the invaders, conciliation became superfluous. As one historian put it, "The Indians were pressed remorselessly when their friendship became of less value than their land." In Virginia, the Indians watched with consternation and alarm as the white men planted tobacco, used up the soil, and every few years moved on to clear new fields. The planters took the Indians' land, first by cajolery and trade, then by force. So swiftly did events move that, within forty years of the founding of Jamestown, the mighty Powhatans were landless and in beggary at the edge of their former homes. Elsewhere the details were different, but white expansion followed the same general pattern.

The barrier of misunderstanding that arose when advancing whites encountered Indians was too high for either people to scale. Some weak and venal chiefs bargained away the rights of their people, but for most tribes the sale of large tracts to the settlers was not a solution to their problems, for they had no land to sell. The warrior chief, Tecumseh, stated the Indian philosophy of nearly all tribes with his reply to the demands of white buyers: "Sell the country? . . . Why not sell the air, the clouds, the great sea?"

To the Indian mind, even after two centuries of acquaintance with the whites, land belonged collectively to the people who used it. The notion of private ownership of land, of land as a commodity to be bought and sold, was still alien to their thinking, and tribe after tribe resisted the idea to the death. Land belonged, they said again and again—in the hills of New York, in the Pennsylvania Alleghenies, and in the Ohio Valley—to their ancestors whose bones were buried in it, to the present generation which used it, and to their children who would inherit it. "The land we live on, our fathers received from God," said the Iroquois Cornplanter to George Washington in 1790, "and they transmitted it to us, for our children, and we cannot part with it. . . . Where is the land on which our children and their children after them are to lie down?"

Had the Indians lacked leaders of integrity, or been less emotionally tied to their hills and valleys, a compromise might have been arranged, but life and land were so intertwined in the Indian scheme of existence that retreat meant surrender of self—and that was unthinkable.

Before the moments of climax came, weaker tribes in all parts of the country made peace, and some of the stronger ones delayed the inevitable by selling parts of their domain. There were fierce chiefs, too, who would not bargain; men repeating the defiance of Canasatego who, representing the Six Nations in Philadelphia in 1742, spoke with contempt of the money and goods acquired in exchange for land. They were gone in a day or an hour, he said, but land was "everlasting."

Yet to many another red man, the new goods had an irresistible allure. Contact with the higher technology of Europeans began to make most of what the Indians had known obsolete, and created needs which they could satisfy only by making increased demands upon the bank of the earth. Once seen, a musket became essential to an Indian warrior; and once an Indian woman had used a steel needle or a woven blanket, she could never again be satisfied with a bone awl or a skin robe. The white man was the only source of the new essentials, and the only way to get them was by trade for things the white man wanted—meat, beaver—and later and farther west, pemmican and buffalo robes. So the Indian, too, became a raider of the American earth, and at the same time was himself raided for his lands by the superior technology and increasingly superior numbers of the white man.

The settlers' demand for new territory was insatiable, and what money could not buy, muskets, deceit, and official ruthlessness could win. Worse, as the bloody thrust and counterthrust went on, hatreds deepened and demagogues argued for a "final" solution of the Indian problem. They coined a slogan that became the byword of the American frontier: "The only good Indian is a dead Indian."

In the westward push, new land became the key to progress, and Indian policy was guided solely by economic expediency. A spokesman for the Ottawa, Sioux, Iowa, Winnebago, and other tribes made this sad and unsuccessful appeal at the Council of Drummond Island in 1816: "The Master of Life has given us lands for the support of our men, women, and children. He has given us fish, deer, buffalo, and every kind of birds and animals for our use. . . . When the Master of Life, or Great Spirit, put us on this land, it was for the purpose of enjoying the use of the animals and fishes, but certain it was never intended that we should sell it or any part thereof which gives us wood, grass and everything."

He got his answer the following year when President James Monroe wrote: "The hunter or savage state requires a greater extent of territory to sustain it than is compatible with the progress and just claims of civilized life . . . and must yield to it."

There was a continent to be redeemed from the wilderness, and the Indians' way of life had to be sacrificed. Thus the policy of forced removal was established, and the Five Civilized Tribes were sent, with scant civility and, in the end, scant humanity, on a thousand-mile "trail of tears" to Oklahoma.

In its latter stages the land war moved into its cruelest phase in California, the Southwest, and the Upper Great Plains. Most of the California Indians were neither as warlike nor as land-conscious as the Eastern tribes. But even this did not spare them, and the most pitiless chapters of the struggle were written by frustrated gold-seekers who organized vigilante raids, killed helpless natives, and subsequently collected from the government for their deeds of slaughter.

After the Civil War the "clear the redskins out" policy approached its dramatic climax. The mounted Indians of the Upper Great Plains and the Apaches of the Southwest were fierce warriors who would not be cornered. It took regiments of trained cavalrymen over twenty years to drive them from their sacred hills and hunting grounds. Out-armed and outmanned, these warriors made fierce counter-attacks, and our American pride was dealt a grim blow when the hundredth anniversary of the Declaration of Independence was interrupted by the news of the Custer massacre. The undeclared racial war did not end until the final tragic chapters were written in the Pyrrhic victory of Sitting Bull at the Little Bighorn, and in the last stands of Crazy Horse, Chief Joseph, and Geronimo.

With the final triumphs of the cavalry, and the uneasy settlement of tribes on reservations, the old slogans gradually disappeared, and the new conscience expressed itself in the saying, "It's cheaper to feed 'em than to fight 'em."

The 1887 Allotment Act, which broke up parts of some reservations and gave individual title to some Indians, further stripped away Indian rights by forcing unprepared tribesmen to deal with unscrupulous land swindlers.

With the passage of time and the steady attrition of old ideas and beliefs, we are at last, hopefully, entering a final phase of the Indian saga. The present generation of Indians accepts the system their fathers could not comprehend. The national government strives to provide the Indian people with adequate health and education programs and to aid them in developing the potential of their human and natural resources. As a singular gesture of atonement, which no civilized country has ever matched, the Congress has established a tribunal, the Indians Claims Commission, through which tribes may be compensated for losses suffered when their lands were forcibly taken from them.

After long years of peace, we now have an opportunity to measure the influence of the Indians and their culture on the American way of life. They have left with us much more than the magic of place names that identify our rivers and forests and cities and mountains. They have made a contribution to our agriculture and to a better understanding of how to live in harmony with the land.

It is ironical that today the conservation movement finds itself turning back to ancient Indian land ideas, to the Indian understanding that we are not outside of nature, but of it. From this wisdom we can learn how to conserve the best parts of our continent.

In recent decades we have slowly come back to some of the truths that the Indians knew from the beginning: that unborn generations have a claim on the land equal to our own; that men need to learn from nature, to keep an ear to the earth, and to replenish their spirits in frequent contacts with animals and wild land. And most important of all, we are recovering a sense of reverence for the land.

But the settlers found the Indians' continent too natural and too wild. Though within a generation that wildness would begin to convert some of their sons, and though reverence for the natural world and its forces would eventually sound in much of our literature, finding its prophets in Thoreau and Muir, those first Europeans, even while looking upon the New World with wonder and hope, were determined to subjugate it.

CHAPTER II

The Birth of a Land Policy:

THOMAS JEFFERSON

The land was ours before we were the land's.

—ROBERT FROST

ON NOVEMBER 11, 1620, after sixty-six days on the stormy North Atlantic, a small ship hove to off the shores of Cape Cod. It is one of the lasting ironies of American history that the shores that were to be a symbol of hope for countless immigrants aroused deep forebodings in those who first came to stay. Nathaniel Morton, the keeper of the records for the Plymouth Colony, observed the face of the land and the faces of the seventy-three men and twenty-nine women aboard, and made this grim notation:

> Besides, what could they see but a hideous and desolate wilderness, full of wilde beasts and wilde men? And what multitudes of them there were, they knew not: for which way soever they turned their eyes (save upward to Heaven) they could have but little solace or content in respect of any outward objects; for summer being ended, all things stand in appearance with a weatherbeaten face, and the whole country, full of woods and thickets, represented a wilde and savage hew.

13

Somehow the entire first phase of New World coloniza-
tion was summed up in this gloomy entry. As the *May-
flower* drew toward shore, probably all but the hardiest of
its passengers were overwhelmed by fear of the wilderness
that faced them. The colonists were ill equipped to pioneer.
Civilization had robbed them of the primitive arts of their
ancestors, and at best they brought with them the limited
skills of city artisans and small farmers. Few knew the
rudiments of fishing, of using native plants for food, of
hunting wild animals, or of building homes in the wilderness. The indispensable colonists were those who knew how
to farm, could catch Indian lore or Indian language on the
run, and had the knack of making peace with the natives.

Yet frightened and ill prepared as they were, the colonists brought with them three things which would assure
their predominance and ultimately change the face of the
continent. First, they brought a new technology. One evening the sun going down over the Appalachians set on an
age of polished stone; the next morning it rose on an age
of iron. From the moment that the settlers won a foothold
and set up their first forge, the sweep of American history
was certain: the Indians would be subjugated; so, too,
would be the land. When the Pilgrims landed at Plymouth
Rock they did not even have a saw, but they brought to
the American continent Iron Age skills that spelled doom
for the Indian way of life. Once the blacksmiths and gun-
smiths set up shop, once the horses and oxen arrived, the ax
and gun and wheel would assert their supremacy.

Second, the colonists brought with them a cast of mind
that made them want to remake the New World. The Indians could plan only from moon to moon, from season to
season, and accepted the world the way they found it, but
the newcomers believed they held their destiny in their
hands and they planned accordingly. The Pilgrims were

men of the Renaissance. Their forebears had developed trial-and-error experimenting into primitive science, and had nurtured the inventor's gift. They knew how to organize as well, and by harnessing work animals to plow and wheel, they could reap where the Indian could not and could sell their surplus in markets overseas.

And, finally, these Europeans brought with them a concept of land ownership wholly different from the Indians': fences and formal papers with wax seals attached were its emblems, and it involved exclusive possession of parcels of land. The European with a title to land owned it whole, no matter whose sweat went into farming it; owned it even if he were a hundred or a thousand miles away. It was his to use or misuse as he saw fit; and he wanted to get and hold as much of it as the law—or the King—would allow.

The influence of the settlers on the land was not as great, at first, as the land's influence on them. As numerous colonies developed along the Atlantic seaboard, the problems and advantages of geography produced different relationships between men and land.

In the South, a warm climate, a wide coastal plain, and rich soil yielded surplus crops to support a leisure class, and led to the development of big plantations and the transplanting of feudal patterns of land ownership. After 1700, the crown and colonial assemblies gave immense grants to men like Lord Fairfax and Lord Granville, on the promise that they would promote settlement. These men led the lives of Old World barons. In 1705, Virginia's historian, Robert Beverly, described them as "men not minding anything but to be masters of great tracts of land—lords of vast territory." The plantations were farmed by sharecroppers and tenant farmers under a system which produced exportable surpluses, but impoverished whole generations of families. Even today, in some areas, nearly all titles (including such landmark estates as Mount Vernon and Monticello) are derived from the great colonial proprietors.

The narrow valleys of New England, however, with their stony soil and severe climate were ill suited to the creation of a plantation economy. This was subsistence-farming country, which could be made productive only through the careful and frugal labors of farmers, and there was little agricultural surplus to harvest for foreign ports or to support a feudal society. New Englanders seemed to gain a kind of rugged independence from the very adversities they faced. The self-reliance later celebrated by Ralph Waldo Emerson was, in part at least, a by-product of these stony New England farms, and a by-product also of the fierce North Atlantic, where the sailors and whalers of Salem, Gloucester, Nantucket, and New Bedford lived and died.

Yet, along with the pattern of New England individualism, there were strong habits of co-operation, arising partly from the necessities of climate and soil, and partly from the fact that the region was settled by members of religious sects with strong group ties. As a result there developed such co-operative institutions as the town meeting, the community woodlot, and the common pasture. Today the Boston Common is the best-known example of a tradition of common land ownership which developed alongside the idea of individual ownership. The town common was, in a sense, the beginning of the public domain.

It was no accident of history that New England leaders called the Continental Congress into session and fired the shots at Concord Bridge, for it was in New England that men had learned to act together while maintaining their individualism; and it was there that the small farmer and the independent landholder got a stake in America.

In the Middle Colonies, where the geography combined

elements of both New England and the South, there were two kinds of landholding, with the large-estate system predominating. Farther from the seacoast, back in the Appalachian valleys, land ownership was modeled more on the New England pattern. There, beyond the domains of the great landlords, the very abundance of land encouraged the ambitions of immigrant commoners to acquire farms of their own, even as the demands of life in the wild hinterlands blurred old distinctions of class and caste. In the face of wolves, savages and blizzards, skill and courage measured men, and nature was the final arbiter of nobility. The hand of London or Charleston or Williamsburg could not reach into the back country; and if a man took up land in the mountains, who was there to stop him or to tell him how to live? The ideas of independence and free land were always inseparable.

Far in the Pacific Southwest another pattern of land use was developing under the aegis of the Franciscan and Jesuit padres. Of all the Europeans who came to America, these men of faith coveted land least. They came with cattle and seeds and saintliness, to build missions and to baptize. Unlike the Appalachian frontiersmen, the padres regarded the Indians as human beings to be civilized rather than as savages to be killed or subdued. In some areas they aided the natives in developing rude irrigation systems and followed the Indian custom of using land as a common asset—a practice which contrasted sharply with the patterns of individual ownership among the Eastern settlers.

Ultimately it was this system of private landholding, fostering a fierce independence of spirit, that was the undoing of the British rulers. They failed to understand that a virgin land settled by men bent on escaping feudal restraints would require a wholly new set of man-land relationships, and new social and political institutions as well.

The frontier expanded and emboldened the thinking of the colonists, and democratic ideas seeped steadily into the dialogue of life. The squatter had no rights against the crown under Anglo-Saxon law, but squatter logic would in due course rule the new continent, overturn ancient laws and customs, and spell out the true meaning of land abundance.

The British failed to reckon the swift pace of American growth or to gauge its influence on the minds of men. Between 1700 and 1776 the population of the British colonies jumped nearly ninefold from 350,000 to 3,000,000, and Benjamin Franklin forecast that within a century there would be more English-speaking people in America than in the British Isles.

Most of these immigrants were land-hungry men. The vast stretches of unused land quickly convinced them that men strong enough to clear a thicket were entitled to own land outright. The mood of the newcomers was voiced by the Scotch-Irish squatters of western Pennsylvania who complained to the British governor that it was "against the laws of God and Nature, that so much land should be idle while so many Christians wanted it to labor on."

However, in the 1760's, King George and his ministers were oblivious to the hopes of the squatters. The biggest British blunder was the Proclamation of 1763, which prohibited settlement beyond the crest of the Alleghenies. As an expedient this order might have made sense, since the British needed time to formulate a plan for settling the Western lands and conciliating the Indians. But the effect of this proclamation was to close the very door that the French and Indian War had just been fought to open. At one stroke it angered the backwoods colonizers, ignored the Western land claims of the seaboard colonies whose original charters had contained "sea to sea" grants, and frustrated the ambitions of the politically powerful land companies that coveted the virgin soil of the Northwest Territory.

The man who, thirteen years later, would write the colonists' answer to this proclamation, was then a seventeen-year-old boy in the back country of Virginia. His name was Thomas Jefferson, and his father was an independent farmer on the far edge of Virginia's plantations. Neither Jefferson nor his father understood the facts of soil fertility. Jefferson's explanation of the practices of Virginia tobacco growers—who exhausted the soil and then moved on—had in it no element of apology: "The indifferent state of agriculture among us does not proceed from a want of knowledge merely; it is from our having such quantities of land to waste as we please. In Europe the object is to make the most of their land, labor being abundant; here it is to make the most of our labor, land being abundant."

But in later years, the mature Jefferson, always open to new ideas, came to see the value of husbandry. All his life he considered himself a farmer by occupation, and the easy habits, once condoned, he came to deplore as his experience broadened. He developed new ideas in horticulture, designed a better plow, imported Merino sheep, and introduced the threshing machine to American farmlands.

To him, agriculture was "the first and most precious of all the arts." By eighteenth-century standards, his sense of husbandry was in the best European tradition, and in his later years he became an advocate of soil studies and crop rotation. Along with contemporaries like John Bartram and Hector St. John de Crevecoeur, he developed European farming practices, and sought to understand the art of renewing the soil for the benefit of future users as well as the current generation. "The land belongs to the living generation," he once wrote. "They may manage it, then, and what proceeds from it, as they please, during their usufruct."

Jefferson's feeling toward the land was one of the strongest influences in the development of his political philosophy. "The small landholders," he wrote, "are the chosen people of God . . . whose breasts he has made his peculiar deposit for substantial and genuine virtue. It is the focus in which he keeps alive that sacred fire which otherwise might escape from the face of the earth."

To a man with these sentiments, the landowning tradition of the feudal aristocrats was an abomination. Jefferson believed that the system of land tenure and distribution adopted would ultimately determine the character of the new society. He saw in America's surplus of resources an opportunity to develop a "natural aristocracy" of talent and virtue. He favored small freehold landownerships which would cause class distinctions to disappear. Growing as Jefferson would have had it grow, this country would have been a rural nation thinly populated by small farmers.

These were his theories, but Thomas Jefferson was more than a theorist; he was a practical reformer, who could go to the heart of a political issue and win others to his opinions. In 1774, when most grievances against the royal reign were directed obliquely at Parliament, Jefferson struck hammer blows at the King and his land policies. He publicly advocated free fifty-acre farms and boldly asserted that lands "within the limits which any particular society has circumscribed around itself, are assumed by that society and subject to *their* allotment." He further counseled his countrymen to "lay this matter before His Majesty and to declare that *he* has no right to grant lands of himself." The land question was the key to a society of equal opportunity, Jefferson was convinced, and the year he wrote the Declaration of Independence he struck a telling blow at the old order by pushing laws through the Virginia Assembly which abolished feudal entails and rights of primogeniture.

The first results of the Revolution—although this is a chapter of history that most Americans have forgotten—

was a program of land reform. The Patriots expropriated and subdivided the Tory estates, and many huge holdings were sold cheaply, given free to small farmers, or parceled out to deserving war veterans. The uncertain claims of many squatters ripened overnight into fee titles that gave the owners the first installment of the new democracy—the right to vote.

The victory at Yorktown, however, did not decide who owned the unoccupied land beyond the Appalachians, and for a time this vexing problem threatened the unity of the Confederation. Settlement had continued during the war years, and seven of the colonies now asserted overlapping claims to the trans-Appalachian country. The crux of the matter was whether these seven colonies should take title to the unoccupied hinterland, or whether it should be a national estate.

Maryland vigorously asserted the case for a national solution by arguing that these lands, "wrested from the common enemy by the blood and treasure of the thirteen States, should be considered as common property. . . ." In the end the Maryland idea won out, and one by one, the colonies ceded over their rights to the lands of the West. Although Virginia's claims were the largest and most legitimate, Governor Thomas Jefferson, in 1781, relinquished them with this observation: "The lands . . . will remain to be occupied by Americans and whether these lands be counted in the members of this or that of the United States will be thought a matter of little moment."

This was our first great land decision, and it was formalized later through Jefferson's work as chairman of the committee of the Confederation Congress, which shaped the historic Northwest Ordinance and the Land Ordinance of 1785. Thereafter, the unoccupied land would be owned by all of the people. The public domain had been created, and a basic ordinance had been enacted that would lead to the establishment of new states. Much later, the public domain would make possible a superb heritage of national forests, parks, and wildlife refuges.

The Northwest Territory, then, was secured for all the citizens of the United States—the vast rich valley, the country of the Shawnees and Cherokees, filled with deer and beaver and rich soil. There was an abundance to this land beyond the mountains that beguiled all Americans. During the early years of the eighteenth century, there grew up a vision of an agrarian paradise that would one day stretch to the western sea. In men's minds the land that lay westward would be the Garden of the World.

Albert Gallatin, Jefferson's Secretary of the Treasury, summed up the universal euphoria with the classic comment: "The happiness of my country arises from the great plenty of land." There was so much of everything—so much land, so much water, so much timber, so many birds and beasts that neither Gallatin nor Jefferson envisioned the day when any natural resources would be depleted. And so, in spite of Jefferson's belief in careful husbandry, his own era saw the beginnings of the Myth of Superabundance that would plunge us headlong into a century of land plunder and land abuse.

In the first years of the Republic, however, our land policies were not designed to get farm tracts into the hands of settlers. The new country had war debts to pay, and Secretary of the Treasury Alexander Hamilton turned to the public domain as a source of revenue. During the early years of the new republic immense areas were sold to land speculators and farm tracts were sold to some individuals.

As a national leader, Jefferson abandoned the idea of free family farms, and supported the Hamilton approach of selling all land for cash. In spite of Hamiltonian policies, the yeoman spirit still walked the hills of the back country, and pressures generated by land-starved men changed the system. The very richness of the land was a force for land reform. The poor squatters demanded, and usually got, pre-emption rights and lower prices, and their counterparts two generations later formed the Free Soil Movement, passed the Homestead Act, and helped fulfill Jefferson's dream of an agrarian empire.

It was Jefferson's basic understanding of the people-land equation, and his confidence in American prowess, that made him a master geopolitician. He understood each maneuver in the imperial power game being played on the chessboard of the American continent by the British, French, and Spanish. In 1802, when yellow fever and Touissant L'Overture's counterattack in the swamps of Santo Domingo crushed Napoleon's plans to restore France's American empire, Jefferson dispatched James Monroe posthaste to Paris to offer a price for "Louisiana." The crucial moment had come, and Jefferson acted swiftly to acquire the heartland of the continent for the American people.

The Louisiana Purchase extended the boundaries of the new nation beyond the rumor of wide rivers, almost beyond imagination. The deal consummated so quickly by Monroe and Talleyrand transferred an area as large as Western Europe for a price of less than three cents an acre. Napoleon himself gave this salute to Jefferson's statesmanship: "This accession of territory consolidates the power of the United States forever, and I have given England a maritime rival who sooner or later will humble her pride."

Timid, literal-minded men cast doubt on the President's legal power to complete the transaction. But Jefferson knew that the Constitution had to grow with the country and his bold assumption of executive power has, over the years, been a ringing reply to those who would limit the role of presidential stewardship. Overnight the American nation was doubled in size, and Manifest Destiny, long before the phrase was invented, had its finest hour.

Jefferson's first move was to send Lewis and Clark to chart this *terra incognita*, but the wisdom of the purchase was vindicated before the explorers made their report. Americans were already on the move. They were not, at first, Jefferson's farmers. For even before events had forced Napoleon to contemplate a sale, a few adventurous men in buckskin had been moving westward through the dark forests of Louisiana.

Boys brought up in the colonies at the edge of the Big Wild were not, after the Revolution, going to stay at home. The tasks of land husbandry would be too tame or too troublesome for many of them, just as the institutions and values brought over from Europe were too binding and too cramped. The new nation, with all its appalling wastefulness, its openhandedness, its generosity and greed, its pride and its independence, was about to begin spreading rapidly from ocean to ocean.

CHAPTER III

The White Indians:

DANIEL BOONE, JED SMITH, AND THE MOUNTAIN MEN

Go play with the towns you have built of blocks,
The towns where you would have bound me!
I sleep in my earth like a tired fox,
And my buffalo have found me.

—STEPHEN VINCENT BENÉT
"The Ballad of William Sycamore" (1790–1871)

"IT WAS on the first of May, in the year 1769, that I resigned my domestic happiness for a time, and left my family and peaceable habitation on the Yadkin River, in North Carolina, to wander through the wilderness of America, in quest of the country of Kentucke. . . ." So ran the words of Daniel Boone's autobiography—and so another chapter in American frontiering began. A tall man with whipcord muscles beneath the buckskin, Boone moved silently through the forest with the soft stride of an Indian. He carried a slim, long-barreled American rifle and it was plain by the way he handled it that it was an extension of his eyes and hands. His animal instincts were honed fine, and in the woods he was sure-footed, with the tireless gait of a man who could lope most of the day if there was good reason—as there sometimes was.

His tradition was older than that of Jefferson's farmers, for he was the essential outrider of settlement. Frontiersmen before him, in quest of their own "Kentuckes," had already traversed more than a third of the continent. Their first great captain, Samuel de Champlain, founded Quebec before Plymouth Rock, and by 1750 his intrepid successors —Joliet, La Salle, Vérendrye, and the *voyageurs*—had paddled and portaged across a great Y extending from Quebec to Lake Winnipeg down to the mouth of the Mississippi. These French and English trapper-explorers were not searching for gold, or for land to farm. Some sought the Northwest Passage; others were after the only treasure their canoes could carry—the finest furs in the world.

Daniel Boone was not a discoverer, in the strict sense. Trappers and Indian traders had penetrated the dark hills before his time, and nearly eighty years earlier La Salle's *courrier de bois*, the peerless Couture, traveled from the Mississippi up the Tennessee River and over the crests to Charles Town on the Atlantic Coast. The trail-opening work of Boone and other hunters in the 1770's struck an auspicious note in our history because it coincided with the events of the Revolution.

The sixth son of a Quaker blacksmith, he was born in Berks County, Pennsylvania, in 1734, nine years before Jefferson. His father later took up a farm in the Yadkin Valley of North Carolina, and here, on the far edge of settlement, Daniel learned the survival code of the frontier and acquired from friendly Cherokees a forest prowess that made him at home in wild country the rest of his life.

While still a young man Daniel went off to war as a

wagoner in Braddock's campaign against the French and Indians, and on returning, he married and carved out a farm of his own, but the urge to have a long look at the undiscovered country got the best of him.

The truth was that Daniel preferred the rifle to the plow, and although word was about that a British Proclamation forbade expansion, the King's rules didn't run in the upcountry, and young men with Indian instincts were bound to crave a westward look.

There was the lure of adventure and the chance to test one's backwoods skills, but there was more than that. At this point in our history, the meaning of the wilderness began to change: the best of the backwoodsmen had mastered Indian woodcraft, and as venturesome hopes dissolved some of the old fears, a new mystique gripped men who lived at the fringe of the frontier. Jefferson sensed it later, and wagered the price of Louisiana on the destiny it held for the American people. It was a mystique known only to men confronted with a virgin continent or an uncharted sea: the undaunted curiosity and quiet fury that led earlier men—Marco Polo, Columbus, Balboa—to take the final chance in their search for the edges of the unknown.

Despite all hazards, men would cross the next river and push through the next gap because certain desires could not be quenched. And so, while young Jefferson formed new ideas about government and the rights of farmers, Daniel Boone left his plow in a half-finished furrow and went into the woods to rediscover an old way of life. Its pleasures were not so cozy as the ones back home, but there were deer and buffalo to kill, and bear to try himself against. There was the dark forest to explore, a game of hide-and-seek with the Shawnees to give a tang to it all, and always the compelling questions: Where does that ridge lead? What lies over the blue hills beyond?

There have been Americans who have had a sixth sense for geography, a map in their heads, and a compass and sextant in their innards. Daniel Boone was one of them, and in 1769 his compass pointed toward Kentucke. He must have fallen in love with the country there, for he wintered over twice and did not return to his family for two years. By the time he came out, he knew more about the bluegrass country than any other white man.

It is hard for us to recreate accurately the life and times of Daniel Boone or to know the Kentucke of his first years. Writing was a skill he lacked, and the autobiography John Filson ghosted for him is two parts Paul Bunyan and one part truth. There were other woodsmen whose achievements at least matched Daniel's—trail blazers like Ben Logan, Colonel James Knox, Simon Kenton, and Michael Stoner—but, thanks mainly to Filson, it was Boone who became the symbol of them all. The book provides more insight into the folk beliefs of the time than into the state of mind of the real Daniel Boone. Filson's Kentucke was a halfway house between the Garden of Eden and the Big Rock Candy Mountain. The soil was richer, the climate was "more temperate and healthy than other settled parts of America"; there were no marshes or swamps; wild game abounded; livestock could roam untended—and manna from heaven could be had for the asking. Filson's tales of Boone, like the legend of Paul Bunyan, helped fill his fellow Americans with optimism that made a paradise of any land to the West.

After we won our independence, the making of land-myths became a national pastime. The myth-makers infected our politics and produced the Go West and Manifest Destiny movements. As long as men were convinced that our continent was a succession of pastures of plenty, they would attempt great and foolhardy deeds, and their forward thrust would ultimately move beyond Jefferson's Louisiana Purchase.

Filson's Kentucke was, in reality, a moving magnet—a neck of the woods that moved a little farther west each year, always one step ahead of settlement. We will never know precisely what Boone saw when he peered down into the valleys of Kentucke from his lookout on top of Big Hill, but we know full well that the Filson-Boone autobiography is one of the early manifestations of the Myth of Superabundance that later caused us to squander our natural resources.

About the time Daniel had his first big look, decisions about the future of Kentucke were being made, and the fever of land speculation involved him in the Transylvania Land Company's scheme to circumvent the King's Proclamation, pre-empt an enormous area beyond the mountains, and plant a new colony in the wilderness.

It was early in the spring when Boone set out with a party of twenty-nine along the wilderness road through the Cumberland Gap. Four weeks later, on the twentieth of April, 1775, the first pack train arrived at the site of Boonesborough. This, unknown to the Kentucke colonists, was a moment of national climax, for just twenty-four hours earlier there had been a beginning of another sort on the village green in Lexington, Massachusetts, where a small band of "embattled farmers" put their future in the hands of the minutemen.

Boone, hunter, explorer, and White Indian, was now an agent of progress and a promoter of towns, but this was an episode he would regret. As a result of his service to the Transylvania Company, he eventually acquired claims on 100,000 acres of choice land. Although he lost much of this land when the Transylvania Company collapsed during the Revolution, he temporarily prospered. But Boone was never at home in a world of fences and farms and legal documents, and as a landlord and land speculator he had a bungling way of letting property slip through his hands.

In 1799, Old Daniel called it quits and headed down-river to accept a Missouri land grant tendered the famous Colonel Boone by the Spanish governor. But fate and his own incompetence in the land business stripped the old man even of this grant. Years later the discouraging story was repeated for the last time when Congress awarded him 850 acres for his "arduous and useful services" to his country. Boone sold this land to pay his Kentucke creditors, and he died in 1820, at eighty-five, a landless freeman still in love with the open country.

Filson made him a rustic George Washington, and put a politician's words in his mouth: Boone considered himself "an instrument ordained to settle the wilderness" and "all his toils and dangers" were made worth while by the prospect of Kentucke becoming "one of the most opulent and powerful states on the continent of North America."

It is far more likely, however, that another utterance of the old man (recalled years later by a grandson) reveals for us the real Daniel Boone: "I had much rather possess a good fowling piece, with two faithful dogs, and traverse the wilderness with one or two friendly Indian companions, in quest of a hoard of buffaloes or deer, than to possess the best township or to fill the first executive office of the state."

Boone the town man was a failure. Boone the folk hero existed only in fiction. It was Boone the outdoorsman who left us a lasting legacy. Land-planning eluded him, but Daniel Boone seemed to hold the notion that every man should have a chance to own a piece of property—to farm, to develop, to use. Implicit in his way of life also was the idea that part of the land should be unowned, or rather publicly owned, as a permanent "hunting ground" for all who like the out-of-doors. His idea of happiness included unspoiled country where the land could sing its authentic songs, and where men could hear the call of wild things and know the precious freedom of the wilderness. By the time Boone died, however, his countrymen were already preparing to dismember the wilderness, and to the east both state and federal governments were disposing of their public lands so rapidly that too little land would be preserved where the young men of the future could relive the adventures of Daniel Boone or know the challenge of wide-open spaces.

Others along the wide Missouri and down the Mississippi knew what Boone meant. One, Mark Twain, would later write a nostalgic story of our early Eden. Huck Finn is a portrait of the American close to the frontier and the wilderness—careless, free of restraints, with none of the unimportant virtues and all of the essential ones. He does not plow or plant or build; he accepts his world with an Indian's casualness and with now and then an Indian's respect. When, at the end, he has a choice of alternatives—to be "civilized" and learn to live up to his newly discovered moral sense, or to stay with nature and to head for the Territories—he picks the latter as the better choice.

The trail of the White Indians did not end with Daniel Boone's burial beside the Missouri. Just a few miles upriver a ten-year-old named Christopher Carson was doing his farm chores and longing to strike out across the plains. In a blacksmith shop in St. Louis, an illiterate adolescent named Jim Bridger was sweating over the red-hot iron and wondering about the high country. In some unknown stretch of forest an educated young man with a Bible in his pack was already moving inevitably toward that day, a year and a half later, which would determine the course of his short life and influence his country's future. The day was March 20, 1822, when Jedediah Strong Smith, aged twenty-two, read this notice in a St. Louis newspaper:

> To enterprising young men. The subscriber wishes to engage one hundred young men to ascend the Missouri River to its source, there to be employed for one, two, or three years.

The man who placed this famous want ad was William H. Ashley, the Lieutenant Governor of Missouri. Within a week he had his takers, and preparations that would revolutionize the fur trade moved forward. The toughest of his recruits would tackle the Indians and the elements, become the most competent outdoorsmen in our history, and write the boldest chapter in the winning of the West.

The magnet that drew the expedition west was beaver, and Ashley's motley band shared a spirit of adventure as expectant and strong-nerved as that which had carried seagoing men around the Cape of Good Hope and across the Atlantic centuries earlier. They had Boone's bent, but the risks they ran were greater and only the lucky ones would live long enough to be town men. Equipped only with horses and rifles and traps, they headed toward the high mountain streams of the Rockies a thousand miles away.

There was a St. Louis saying that God always stayed on his own side of the Missouri, but raw nerve was the long suit of these "enterprising young men," and Bernard De Voto has given us a superb description of the Spartan demands of this frontier:

> The frontiersman's craft reached its maximum and a new loneliness was added to the American soul. The nation had had two symbols of solitude, the forest and the prairies; now it had a third, the mountains. This was the arid country, the

land of little rain; the Americans had not known drouth. It was the dead country; they had known only fecundity. It was the open country; they had moved through the forests, past the oak openings to the high prairie grass. It was the country of intense sun; they had always had shade to hide in. The wilderness they had crossed had been a passive wilderness, its ferocity without passion and only loosed when one blundered; but this was an aggressive wilderness, its ferocity came out to meet you and the conditions of survival required a whole new technique. . . . In that earlier wilderness, a week's travel, or two weeks' travel, would always bring you to where this year's huts were going up, but in the new country a white man's face was three months' travel, or six months', or a year away. Finally this was the country of the Plains Indians, horse Indians, nomads, buffalo hunters, the most skillful, the most relentless, and the most savage on the continent. . . . Mountain craft was a technological adaptation to these hazards.

The decade of the 1820's was the golden era of the fur trade and gave birth to the free-trapper tradition. Out of that era have come as many legends as facts, and the legend-makers have bequeathed us, larger than life, a Kit Carson and a Jim Bridger. But most of the mountain men never reached the glory road of Western fiction: they did not find their Filson, or they ventured too much, or their luck ran out. This is not their story, but we owe it to them in passing to recite names like Tom Fitzpatrick, William Sublette, Antoine Leroux, "Black" Harris, the Bent brothers, Manuel Lisa, and Étienne Provost.

They were mostly unmarried, and no umbilical cord tied them to a farm or family. Land ownership never entered their minds, for these men were the complete sons of the wilderness, the true White Indians. They shared Boone's illiteracy and stoicism; and their dreams of wealth, if they ever had any, were as foredoomed as Daniel's. Their business was the killing of beaver and De Voto rightly called it "as ruthless a commerce as any in human history."

Francis Parkman, the historian, was not far wrong when he referred to them as "half-savage men." They had to be to survive, and they could move safely through the high country precisely because their language and dress and sometime bedmates were Indian. Among them were the managers who led the parties, and arranged for the marketing of pelts in St. Louis or Santa Fe. The most remarkable of these men was surely Jedediah Smith.

At the age of twenty-three Jed Smith went upriver with Ashley, and ten years later he was lanced to death by the Comanches on the Cimarron. In the interim he scouted nearly every major stream west of the Mississippi, survived the three worst massacres of the Rocky Mountain fur trade, and compiled a lists of firsts that stands by itself. In his quest for untrapped beaver he traveled farther and saw more of the West than any of his contemporaries—including Lewis and Clark. Had he taken the time to put his dead-reckoning knowledge of topography on paper, American map-making would have jumped twenty years ahead.

Physical courage was the *sine qua non* of mountain men, and Smith proved his at the Arikara massacre, which made Ashley's first expedition a failure. A grizzly marked Smith's face for life a few months later, and his endurance and personal force carried him within three years from head of a hunting party to senior partner of the dominant fur-harvesting partnership in the Rockies—Smith, Jackson, and Sublette.

Smith roamed the tributaries of the Upper Missouri, went as far as the Bitterroot, and later traversed the South Pass gateway into the Great Basin.

But his most astonishing overland odyssey began in the late summer of 1826 when, in search of new beaver country, he headed southwest from the summer rendezvous at Bear Lake. In a year's time he pushed to the lower Colorado River, through the parched Mojave Desert to the California missions (where he got a reception befitting the first American overland party), trapped his way up the San Joaquin Valley, made the initial crossing of the Sierra Nevada, and returned across the Great Basin to his headquarters. As if this weren't enough for any man, he set out a month later to retrace his steps, lost half of his men to a savage Mojave attack, and blazed the first trail from southern California north to Fort Vancouver on the Columbia River. Of twenty men who had survived the Mojave attack only four lived through an ambush by the Umpqua Indians in southern Oregon, but Smith pushed on up the Columbia to Fort Colville and two years after his departure rejoined his partners at Pierre's Hole in southern Idaho. Unlike the military expedition of Lewis and Clark, Smith's men were ill equipped, and he imposed discipline only by the force of his own fortitude. Before he was through, his travels covered nearly three times the distance spanned by Lewis and Clark in their earlier voyage of discovery.

The objective of these expeditions was beaver, but Smith once wrote in his journal that he was also led on by "the love of novelty common to all." Weather-beaten and raw-hide-tough, he had appeared in California and Oregon, without credentials or a flag, but the very presence of this scout in buckskin was a rude announcement of Manifest Destiny to the colonial outposts of Britain and Spain.

None of the mountain men got rich trapping, and most died poor. Beaver plews sold for six dollars apiece in peak years, and a good trapper could make one thousand dollars a season. But at the summer rendezvous the fur companies charged outrageous prices for supplies hauled in from St. Louis, and most of the time the trappers decided to stay on another year in the high country and hope for a bumper harvest. A few cleaned up, and John Jacob Astor, running part of the show from back East, became the richest man in America because he knew how to organize the extermination of the beaver. But while in pursuit of beaver, Smith and the mountain men planted outposts, learned a way of life, and enjoyed a once-upon-a-continent freedom to explore some of the finest mountain country in the world.

Some of them later used their hard-won knowledge of the wilderness to guide the wagon trains that began to spill out across the plains in the '40's and '50's. The last of the White Indians gave what discipline they could to an undisciplined migration, found the waterholes, and kept the emigrant-trains moving.

What is the land legacy of the mountain men? Legends aside, surely we owe them a larger debt than we have yet acknowledged.

Thomas Jefferson knew their fathers, and wagered all on the ability of such men to respond to the wilderness challenge. For nearly a generation these men were the American presence in an area where sovereignties overlapped and national boundaries were still undefined. As Jefferson foresaw, in the last analysis it would be unafraid men, outward bound to conquer and explore, who would fill the vacuums and fix the boundaries of the nation. The mountain men, if they failed in all else, made him a prophet.

Like Kit Carson, the best of them were "cougar all the way," and they established an ideal of prowess which entered the marrow of our national character, which saw us up San Juan Hill and through two world wars. Our sentimental fondness and genuine respect for this ideal is, I suspect, the secret of the durability and fascination of the American "Western."

But all qualities, including a mustang human spirit, have their defects, and these, too, must be entered in the record. The trappers' raid on the beaver was a harbinger of things to come. Their undisciplined creed of reckless individualism became the code of those who later used a higher technology to raid our resources systematically. The spiritual sons of the mountain men were the men of the next wave—the skin-and-scoot market hunters, the cut-and-get-out lumbermen, the cattle barons whose herds grazed the plains bare.

It is neither fair nor quite true to say that the tradition of thoughtless land exploitation started with the mountain men, but certainly a part of it can be traced to them. Leatherstocking, James Fenimore Cooper's idealized frontiersman, found God in the trees and water and the breath of summer air; but the true-life mountain man made his demands on America's abundance without thought, without thanks, and without veneration for living things. These men embodied, as few others have, one facet of the self-reliance of which Emerson later wrote, but they wholly lacked the self-discipline which alone could give it grace and meaning.

In all this, the circular process of history was at work. The land was determining the character of men, who, in turn, were determining the future of the land itself. The result of this interaction was the clearest possible example of the American ambivalence toward the land that continues to dominate our relationship to the continent and its resources. It is a combination of a love for the land and the practical urge to exploit it shortsightedly for profit.

It is in their love of the land that the frontiersmen and the mountain men have given us a lasting gift. Each new generation of Americans is inspired by their ideal of individual prowess. In our few remaining wild lands we can still catch a glimpse of the world of Kit Carson and Jim Bridger and Jed Smith—the world that shaped our character and influenced our history. The spirit of Boone and the mountain men still walks the woods and Western ranges. A stanza Vachel Lindsay once wrote is their ultimate epitaph:

> *When Daniel Boone goes by, at night,*
> *The phantom deer arise*
> *And all lost, wild America*
> *Is burning in their eyes.*

CHAPTER IV

The Stir of Conscience:

THOREAU AND
THE NATURALISTS

*I believe a leaf of grass is no less than the journeywork of
the stars,
And the pismire is equally perfect, and a grain of sand,
and the egg of the wren,
And the tree-toad is a chef-d'oeuvre for the highest,
And the running blackberry would adorn the parlors of
heaven,
And the narrowest hinge in my hand puts to scorn all
machinery,
And the cow crunching with depress'd head surpasses any
statue,
And a mouse is miracle enough to stagger sextillions of
infidels!*

—WALT WHITMAN "Song of Myself"

THERE WAS another kind of moccasin stalking the wilderness at the end of the eighteenth and through the nineteenth century. Its wearer was not Indian, though he quickly grasped the age-old insights of Indian living. Nor was he White Indian: the brain was too self-aware, the mind too attuned to overtones, the inner ear too acute. This stealthy foot belonged to a new breed of hunter, the naturalist, who sought the miracle of nature as the Red Indian sought a deer or the White Indian a westward pass.

The interests and personalities of such naturalists as Bartram, Audubon, Parkman, Emerson, and Thoreau were as varied as their origins: one was a pre-eminent philosopher, one a pre-eminent historian, and the others were regarded as more quaint than profound by their contemporaries.

But in different ways these five were land-conscious men who owed a debt to the Old World and shared a desire for fresh insights into the nature-man equation. Likewise, they shared a deep interest in, and respect for, America's aborigines, and they rejected the common notion that the European emigrants had nothing to learn from the natives.

Some were basically pastoral men (although, by Jed Smith's standards, most of them were greenhorns and backyard bird watchers) ; others were more at home in parlors; but as a group they were bent on seeking out the larger meaning of nature and making it part of the woof of life.

Only two of them were intimates, and the paths of some never crossed, but they shared a virgin continent as a common laboratory, and viewed it with an eye of discovery that probed beyond the obvious. In a much different way they were as individualistic as the mountain men, and each contributed to new currents of thought that reshaped our thinking about the American land.

It began, perhaps with the Bartrams. Daniel Boone, Thomas Jefferson, and William Bartram, naturalist, were born within ten years—and died within six—of each other. While Jefferson stood on the portico of Monticello contemplating the agrarian advance, while Boone shaded his eyes at the summit of Cumberland Gap, William Bartram was down in a valley somewhere, on shank's mare, inspecting palmated chestnut leaves, bird nests, vines, and berries. Jefferson knew Europe, and spent a lifetime borrowing its sophistications; Boone had shed Europe as a spring snake sheds its old skin; William Bartram depended upon Europe for his bread and butter, and his self-appointed task was to win respect for the Old World's new art of nature study.

Bartram's father, John, once the King's botanist, had created America's first botanical garden on the banks of the Schuylkill River. After one five-week jaunt of over 1,100 miles, John complained that no one would "bear the fatigue to accompany me in my peregrinations." He was elated when his son William acquired his thirst to witness and describe every facet of nature. William, whose long stride matched his inquisitive eye, wandered thousands of unfenced miles on the eastern side of the Appalachians, and his journals bubbled over with the richness of a countryside teeming with "frogs in springly places," "gay, vociferous and tuneful birds," "myriads of fish, of the greatest variety and delicacy, sporting in the crystalline floods," and "Elysian springs and aromatic groves."

He wrote so eloquently that his *Travels*, published in 1791, received high praise in Europe. Carlyle wrote Emerson that all libraries should have "that kind of book . . . as a kind of future biblical article." Chateaubriand in France, and Wordsworth and Coleridge in England, used the *Travels* as a basic wild-land reference book. Jefferson turned to Bartram when he needed strawberries and other plants for Monticello, and years later asked him to join Lewis and Clark's expedition, but William, old and walked-out, regretfully declined.

As a self-taught disciple of Professor Linnaeus, William Bartram made botany popular, and he broadcast abroad— through his writings and exchanges of plants and seeds—

the wonder of the American scene. Science and institutions like the Smithsonian had a chance to grow in the United States once his work was under way, and the generation of nature writers and nature students which followed had a frame of reference from which it could measure its own insights.

Of Jefferson's contemporaries, none achieved more popular fame than Haitian-born Jean Rabin, the naturalist whose name evolved to John James Audubon. Unlike his modern followers, who hunt with binoculars, Audubon took pleasure in shooting birds in order to identify them, and he chose the best for painting. In Florida he poked through the bayous and keys and boasted of shooting enough birds to make a feathered pile the size of "a small haycock" in a single day. From 1820 to 1826, Audubon hunted species after species, securing specimens and painting them with regal distinction for his *Birds of America* books.

His bird paintings were no sooner completed in 1839 than he began an ambitious work on North American quadrupeds and made a trek far up the Missouri in 1843 to gather facts firsthand. For two months Audubon was a guest at Fort Union at the mouth of the Yellowstone. He rode with buffalo hunters, thrilled to the sound of "wolves howling, and bulls roaring, just like the long-continued roll of a hundred drums, "and commissioned trappers to bring him odd and unusual species. Passing Indians yelled with delighted recognition at the drawings he showed them from his portfolio, and he sketched night and day.

Audubon began the "Quadrupeds" collection at a time when it was impossible for a single naturalist to encompass all the fauna of the country. But he was indefatigable and saw his plan fulfilled after the Smithsonian was founded in 1846 and government survey parties shipped ponderous collections to Washington for analysis and classification.

Though Audubon never seems to have regretted his own big kills, in his later years he lashed out at the reckless raids on wildlife. He lamented the disappearance of deer in the East and denounced the "eggers of Labrador," who slaughtered sea birds for fish bait. On witnessing the depredations of one fur company killing large numbers of mink and marten, he cried out "Where can I go now, and visit nature undisturbed?"

Audubon's work heightened American interest in nature, and stirred the protective instincts of sensitive men. He was not a moralizer, nor did he seek to organize a crusade, but he became a symbolic figure in the fight for the preservation of wildlife, and after his death the society organized in his name set up a continuing protest against the slaughter he abhorred.

Audubon was a link between the mountain men and the naturalist-philosophers. Like the former, he was primarily a man of action rather than a prophet or profound thinker. Like the latter, he took delight in the systematic observation of wildlife and considered nature to be an object of study, not of conquest. His work is a manifestation of that same bedazzled love of the American scene that turned up in John Filson's *Kentucke* and the works of William Bartram. Audubon did not merely record his creatures; he endowed them with his own enthusiasm. The best of his birds not only reflect their own beauty, but are alive with his excitement.

Three years after Audubon's foray up the Missouri, amid the wagons-west tide of 1846, Wyoming was host to another Easterner—a Boston-bred collegian of twenty-three, Francis Parkman—who was to become America's foremost interpreter of the encounter between Iron Age men and a virgin continent peopled with powerful native tribes. Wrapped up in his complicated life and character are all the contradictions and attractions of civilization and wilderness and their tragic, inevitable collision.

Francis Parkman, in his own words, was a man "haunted by wilderness images day and night." This fact alone is an indication of how deeply the wilderness experience was beginning to invade the American consciousness. In background he shared nothing at all with outdoorsmen like Boone or Jed Smith. He had grown up among the merchant princes of Beacon Hill in the hub of American civilization, played beneath the benign eyes of portraits of Puritan and Revolutionary ancestors, and was clearly destined, from an early age, to enter the academic halls of Harvard College.

His life might have been rigidly conventional had he not found a wonderland of rocks, gorges, and pine woods near his grandfather's farm when he was eight years old. This glacier-hollowed dell, ignored by colonial farmers, bore the intriguing name of Middlesex Fells. Here a protected city boy could shuck off his shoes and ramble around, could trap woodchucks and shoot at birds with an arrow, could poke his nose and fingers into all sorts of rocky recesses, and could put some timber in his spine. Francis had a second cousin, Henry Adams, who described the contrast of city and country as "the most decisive force he ever knew." So it was for Francis, who learned all that the Fells had to teach him and later put it to good use.

In his college years he ventured into the more sizable wilderness of the Magalloway River, where he and a fellow sophomore made a canoe exploration, "seeking a superior barbarism, a superior solitude, and the potent charm of the unknown." The Magalloway was a miniature Wild West, and it fixed Parkman's attention on the great New World drama of men and land. There he saw his first Indian, first slept under the stars, and first shot big game.

Soon after, he obtained letters of safeguard from the American Fur Company and headed West on the Oregon Trail to find some Sioux. The Brahmin held himself aloof from the emigrants in covered wagons along the trail, for "this strange migration" of zealots and land-hungry farmers perplexed him. To his Boston tastes, the emigrants were unkempt, homespun people, oblivious to the magnificence of the country. He saved his admiration for the old frontiersmen who seemed to have the breath of the wilderness in them.

The climax of Parkman's trip came when he was accepted by a band of Oglala Sioux, who were roaming the high plains with primitive vigor. Sick from alkali water and the rigors of his trip, Parkman stayed at the side of Chief Big Crow seventeen days as his tribe hunted buffalo and prepared for war. This Oglala interlude, at twenty-three, was the summit of Parkman's active life. The simple ways of the Sioux and their attachment to the land reverberated in his mind as long as he lived.

It was surely some sixth sense that had sent Francis Parkman to the wilderness in 1846 before his body failed him. He saw and felt the open land before it was too late and met the rival forces that were contesting for dominion. The following winter he was felled by a nervous breakdown and, a half-blind invalid, turned his imagination inward and began his life's work.

The darkness was closing in as he sat at home in Boston and wrote of his frontier adventures. He wrote blindfolded with the aid of a wire frame, but he could reach out and touch the buckskin, feathers, arrows, and lance of the Oglala and remember the exaltation of his Indian encounter. The result was his first book, *The Oregon Trail*, published in 1849.

The abiding significance of Parkman's wilderness research unfolded in the forty-five years after his Oregon journey, as he wrote eight volumes of history. In his books he savored the excitement of the wilderness, yet his work eventually became a vast inquiry into geography and anthropology, a study of a virgin land and the competing cultures that struggled to master it. Parkman's histories abounded in glowing descriptions of wild country: he was an avid reader of the earlier naturalists and once pirated passages from Bartram's *Travels* to recreate the pristine aspect of the Florida's savannas, but such words were mere background music, for once Parkman's organ-toned prose began to flow, the land had found its historian. His language caught the melody of the red man and the Jesuits, of Pontiac and Frontenac, of Wolfe and Montcalm, and occasionally one even caught a glimpse of Parkman as he revealed himself through his word pictures of the shy priest, Robert Cavelier, Sieur de La Salle.

Parkman wrote with the style of a novelist and the insight of a historian; to him the saga of American settlement had all the overtones and grandeur of classic tragedy. The European invaders, Parkman could see, would inevitably subjugate their Stone Age adversaries. Indian ways and Indian values would be eradicated by the onrush of a "superior civilization." Here were the grand themes of Parkman's books, and his contribution as a historian rests on his feeling for the land and his understanding of the irrepressible conflict over its future. In his last years, he took up his pen for the forests and for the Indians, and he died a city man who loved wild things, a sick man who esteemed physical prowess and saw that unspoiled nature had indispensable truths to tell.

While Parkman was a boy, preparing to enter Harvard, not far away a movement was developing that invigorated and reoriented American thought. It began in Concord and its captain was Ralph Waldo Emerson, who bade his countrymen to drop their imitation of Old World ways and strike out on their own.

To Emerson the nature discoveries of such men as Audubon and Bartram were valuable, but no substitute for his own discoveries. He was concerned not merely with the details that preoccupied the specialists, but also with developing some broad conclusions about man and his environment. To do so it was not enough to rely on books and secondhand experiences; every man should be his own naturalist and his own philosopher.

The principal thesis of his essay, *Nature*, published in 1836, was that the individual should "enjoy an original relation to the universe," and that message become the cardinal article of faith in the philosophy of New England transcendentalism. "The inevitable mark of wisdom," he advised, "is to see the miraculous in the common."

To pursue his vision more intently, Emerson steeped himself in Plato, Goethe, and fresh air. The easiest way to develop Olympian insights was to turn the mind into an aeolian harp and attune it to the winds and sounds and rhythms of nature. Many of Emerson's essays were forest prose poems and, for our purposes, the principal significance of the transcendental movement lies in the fact that it is rooted in nature. Solitude and meditation were Emerson's meat and drink, and the inner harmonies of life were clearest in his mind in the out-of-doors.

Emerson had many messages for his countrymen, but none was more profound than his conviction that they would find their own pathway only if they gave up their veneration of the Old World, cultivated self-reliance, and responded to the rhythms of the American earth:

> Embosomed for a season in nature, whose floods of life stream around and through us, and invite us, by the powers they supply, to action proportioned to nature, why should we

grope among the dry bones of the past . . .? In the woods is perpetual youth. . . . In the woods, we return to reason and faith. There I feel that nothing can befall me in life,—no disgrace, no calamity . . . which nature cannot repair . . . the currents of the Universal Being circulate through me; I am part or parcel of God.

The transcendental philosophy needed poets to sing of the natural world and natural men, so Emerson wrote verse and was the first to hail the roughhewn images of young Walt Whitman. Before Emerson's work was done, he became the first great American philosopher. His message was an affirmation of optimism, but in terms of land the optimism was the mote that marred his vision. Self-reliance was a quality that had defects, and already men who possessed it to excess—men who were solely concerned with immediate profits—were plundering their way through the forests and across the countryside.

Emerson, however, viewed such developments with unconcern. He was so attached to his hopes for America that he dismissed flagrant waste with the euphoric observation that pirates and rebels were the real fathers of colonial settlement, and men would adopt sound policies once the frontier was settled and the ennobling influence of nature took effect. This was sophistry, as Emerson would have realized had he roamed the back country as Bartram did, or traveled west with Parkman.

If Emerson made no major protest against resource waste during his lifetime, his grand themes nevertheless helped arouse interest in the natural world, and inspired his Concord neighbor, Henry David Thoreau.

Fourteen years Emerson's junior, Henry, in 1837, gave a Harvard commencement address in which he enlarged on Emerson's *Nature* essay and offered the proposition that the order of things should be reversed: the seventh day should be a day of work, for sweat and toil; the remaining six days man should be free to feed his soul with "sublime revelations of nature." He then proceeded straightaway to turn his life into an object lesson of this expansive proposal.

Thoreau was generally a most ungregarious fellow. Although intimate with Emerson and a few friends, he was a naturalist's naturalist: he had animal sense, was self-centered to a marked degree, and had the leisure and single-mindedness of a bachelor with only tangential ties to hearth and home. Henry's abstemious traits and Old Testament ways made it difficult for most men to understand him, and his porcupine personality kept even his friends at arm's length. James Russell Lowell tried to type him as an eccentric bent on escaping from the real world, and most of his neighbors thought him a rather harmless handy man who spent much too much time wandering in the woods.

Thoreau's quarrel with men of Lowell's frame of mind, however, was over the identity of the real world. He rejected the idea that the affairs of men should be centered on getting and spending, just as he looked with misgivings on the growing domination of the landscape by the engines and establishments of the industrial revolution. Thoreau was critical of a "blind and unmanly love of wealth," and he was alarmed that men were losing their ties to the land.

What would it profit, he asked himself, if men gained a whole continent, but in the process lost contact with the wellsprings of human renewal? It was painfully clear to Henry that most men were using their best energies on "business" and neglecting the business of living each day to the fullest. The proper study of mankind was nature, and there was a higher law than "progress."

This estrangement from nature, this withering of the senses, became Thoreau's preoccupation. Concord was too

worldly for him, and in 1845 he repaired to a cabin at
Walden Pond to set up a "laboratory" and conduct a simple experiment in living. In order to carry out his self-
appointed assignment as "inspector of snowstorms" and
surveyor of all outdoors, Thoreau minimized his ties with
town life, and maximized his contacts with nature. His
touchstones were simplicity and economy.

Both Thoreau and his followers have idealized Walden,
but by any standards it was unique, for it was a here-and-
now utopia and its prime object lesson was that heaven lay
all around if men would only open their eyes to it. Those
who did so would surely conclude that "those are richest
whose pleasures are cheapest." Keeping an orchard between himself and the nearest farmer, Thoreau carried on
intimate discourse with animals and birds, followed the life
cycles of plants, and measured ponds and rivers as he strove
to catch the quintessence of the living world.

After twenty-six months, Thoreau left his Walden cabin
for good, but save for two trips to "strange country"—one
to Cape Cod, the other to the woods and rivers near Mt.
Katahdin in northern Maine—he continued to root his
transcendental experience in the Concord countryside.
Filled with the conviction that even a man who diligently
applied himself could not really *know* more than a six-
mile-square plot, he described every aspect of Concord's
rivers, inquiring into what the fishes and insects ate and
where they died, examining its stones and depths, its loons
and ospreys. Nor was Henry enticed by the suggestion that
he visit England and Rome. His scorn for such travel was
his way of saying that each man should take his insights
from his native ground.

Thoreau was so intent on exploring the depths of his
own fields and streams that the frontier did not excite his
interest, and he urged: "Be rather the Lewis and Clark and

Frobisher of your own streams and oceans; explore your
own higher latitudes . . . explore the private sea, the
Atlantic and Pacific Ocean of one's being alone."

Thoreau developed for the land a more explicit and all-
consuming reverence than his predecessors. When he
helped Louis Agassiz collect turtles for laboratory use, he
noted in his journal that it was a "murderer's experience
to a degree." Firearms were not for him, and his role, in
the only deer hunt in which he participated, was that of
"chaplain and conscientious objector."

Thoreau was one of our first preservationists. He believed that some landscapes should be left unspoiled and
that animals should have inviolate places of refuge. In
1858, more than a decade before Congress set aside Yellowstone as the first national park, Henry Thoreau made
a plea for "national preserves, in which the bear, and the
panther, and some even of the hunter race may still exist,
and not be civilized off the face of the earth—not for idle
sport or food, but for inspiration and our own true recreation."

He protested destruction of landscapes and wildlife. Appalled by the work of dam builders on the Concord, he
wrote an impassioned appeal for the shad: "Poor shad,
where is thy redress . . . who hears the fishes when they
cry?" And on viewing the forest waste in the logging
camps of Maine, he exclaimed: "It is the poet . . . who
makes truest use of the pine. Every creature is better alive
than dead, men and moose, and pine trees, and he who
understands it aright will rather preserve its life than
destroy it."

It is significant that most of these words of dissent were
recorded privately in Thoreau's journal and not printed for
public consumption. At a time when the raid on resources
was gathering momentum in the forests of Wisconsin, the

mountains of Colorado, and the valleys of California, Thoreau was concerned primarily with philosophy rather than public action. Politicians who tried to influence public opinion were, he thought, in a "narrow field, and a still narrower highway yonder leads to it. I am pleased to see how little space they occupy in the landscape."

Unlike Emerson, Henry was mildly alarmed by the raider spirit abroad, but he failed to realize that the land spoilers were already in command, and were committed to a course of action that would destroy the land values he prized the most. With his negative feelings about government and politics, he failed to perceive that it would take government action to stop the destruction. There were other contradictions: although he abhorred the very thought of social action, land conservation could not begin until men organized for action; he was antireformer, but it would take the crusading zeal of reform-minded men to save the woods and wildlife; he was, moreover, the most thoroughgoing nonconformist alive, though the dangerous drift of the time pointed to the need for conformity to minimum rules of resource management. In short, government action was necessary to curb the exploitation of resources and allow the land to renew itself, but Henry David Thoreau was constitutionally and unalterably antiprogram and antigovernment.

Yet if Thoreau "insisted on nibbling his asparagus at the wrong end," as Oliver Wendell Holmes charged from the parlors of Boston, it was for an understandable reason. Ideas must precede action, and sometimes the seeds of thought have a long period of germination. Reform must begin in the minds of men, and a significant segment of society would have to accept the Walden values before effective action could commence.

Men like Thoreau, Emerson, Parkman, Bartram, and

Audubon were the idea makers, the essential forerunners of the conservation movement. They started new processes of thought; they began the development of an American land-consciousness and set in motion a salutary countercurrent of ideas against the raider spirit of their era. These men saw the cosmic in the commonplace, and sought to grasp the whole of existence by acquiring fresh insights into everyday life. Although some of their writings were long ignored, the twentieth century eventually rediscovered them.

Meanwhile, beyond the confines of Concord, forests were being mowed down, wildlife was disappearing, and soil was being washed away. Paradoxically, the men who were beginning to raid the continent were motivated by an undisciplined extension of Thoreau's own individualism. They shared his contempt for politics and public men, and they carried his every-man-for-himself attitude into the mountains and the market place, making irresponsible independence a creed which rejected wise management of resources.

It would soon be apparent that the vision of Thoreau and Emerson was ahead of its times. The Concord transcendentalists, with their message of reverence for the land, would not be heeded until crisis lay across the nation's path like a fallen tree.

CHAPTER V

The Raid on Resources

*It was all prices to them: they never looked at it:
why should they look at the land? they were
Empire Builders: it was all in the bid and the
asked and the ink on their books. . . .*

—ARCHIBALD MACLEISH
"Wildwest"

THE Big Raid on resources began, in a sense, with
mountain men and their beaver traps, and reached a series
of high points in the last decades of the nineteenth century.
Its first phase involved only the harm caused by primitive
tools, but the second was linked to the machines of the
industrial revolution, which made possible large-scale harvesting of resources—and large-scale land damage.

It was the intoxicating profusion of the American continent which induced a state of mind that made waste and
plunder inevitable. A temperate continent, rich in soils and
minerals and forests and wildlife, enticed men to think in
terms of infinity rather than facts, and produced an over-
riding fallacy that was nearly our undoing—the Myth of
Superabundance. According to the myth, our resources
were inexhaustible. It was an assumption that made wise
management of the land and provident husbandry superfluous.

A growing nation needed wood for housing and fuel and shipbuilding, and the biggest of the Big Raids began in the woods. The virgin forests of North America were among the masterpieces of the natural world: east of the Great Plains nearly every acre was covered by trees; to the west softwood stands flourished on the slopes and in the valleys of the Rocky Mountains; and rising above the Pacific shore line, in the most productive timber zone in the world, redwood and fir stands provided a crescendo of arboreal splendor.

Europe had hardly a dozen tree types. The American expanse had more than a hundred, and our many soils and climate zones produced the largest and oldest trees, and the most accessible commercial stocks on any continent. The first task of forest-bound colonists was to develop woodsmanship: homes and stockades had to be roughhewn, land cleared, and firewood cut. Farming awaited the work of the broadax: clearings could be carved out of the virgin thickets only through great effort or by the deliberate use of fire.

Tree cutters were the advance men of agrarianism, and the worst acts of forest destruction were oftentimes explained away with the carefree rationalization that such devastation was necessary to "let daylight into the swamps." The common assumption was that trees, like Indians, were an obstacle to settlement, and the woodsmen were therefore pioneers of progress.

Until the machines came along, the Ax Age American and the Stone Age Indian could inflict only limited damage. Small local sawmills provided all the finished lumber needed in the early days, and it was about 1800 before cities required lumber in wholesale lots, and enterprisers learned to organize machines and men to produce it. Then came the circular saw and the steam mill. The wood-pulp process for making paper was invented, and a frantic wood rush began that would strip most of our forests and help puncture the Myth of Superabundance. The sad truth is that when the big lumber raids got under way, our thinking on forests was fixed, and we regarded those who invaded the forest heartlands as the leaders of the westward advance.

Lumbering quickly became our largest manufacturing industry. Conditions could not have been more favorable: timberland was cheap, labor was cheap, and a small investment would outfit a mill and start a log boom. The onslaught began in New England, moved across New York and Pennsylvania, leveled the vast pineries of the Lake states, and finally cut a wide swath through the yellow-pine stands of the South and the West. Until 1909, the year output reached its peak, production continued to increase in crazy cycles of boom and bust. There was enough wood for a thousand years, the optimists said, but the lumbermen leveled most of the forests in a hundred.

The devastation was not caused by logging alone: careless loggers caused fires that burned as much as 25,000,000 acres each year. Some irresponsible lumbermen deliberately set fire to the debris they left behind, thus destroying seedlings that would have replenished the ravaged forests.

The average life of a sawmill was the twenty years it took to strip a hinterland. As this brawling industry ran its course, such towns as Bangor, Albany, Williamsport, Saginaw, Muskegon, Eureka, and Portland in succession briefly boasted the title "lumber capital of the world."

Lumbering, in its raider phase, was a strip-and-run business: the waste of wood was enormous, and when the best stands had been cut, the operator dismantled his mill and moved it farther west, letting the raped land go back into public ownership because the taxes went unpaid. In some areas sawmill ghost towns still clutter the landscape. Trees, like gold or silver, were "mined," for the land-skinners wanted the quickest profits the system would allow.

It was an era, too, of human heroics, and the Paul Bunyan loggers and lumberjacks who took on and tamed the howling wilderness had a prowess that still commands our respect. Their feats of main strength, however, somehow symbolized the mindless, planless approach that ruined vast sections of our land.

The destruction of the American forests was facilitated by the Great Giveaway of land, which made the raids possible; by the optimism of the people, which made the Great Giveaway plausible as a policy; and by the political power of the raider captains themselves, which kept exploitation going full tilt once the harvest began.

Many of the states outdid the federal government in "giving away" timberland. Maine and Pennsylvania sold off enormous tracts for twelve and one-half cents an acre. North Carolina, not to be outdone, auctioned off choice hardwood stands for ten cents. Some of the best federal timberlands were also put on the market at bargain prices. "Sell cheap" was our slogan, and leaders like President Andrew Jackson and John C. Calhoun wanted to cede all federal land to the states free of charge and be done with national trusteeship.

For shrewd lumbermen, however, there were ways to take trees without buying huge tracts of land. A common practice was to purchase adjacent forty-acre pieces of public land and cut all around them. The "round-forty" was the classic joke of the lumber industry.

A few provident grants were made, but "giveaway" was the rule, and prodigality reached its peak between 1850 and 1871 when an area larger than France, England, Scotland, and Wales was granted to the railroad companies.

The loose forest "homestead" laws passed by Congress in the 70's and 80's prompted the forest historian, Samuel Trask Dana, to conclude that by 1885 "ninety percent of the entries under the Timber Culture Acts were fraudulent."

The politics-ridden General Land Office was given neither the men nor money needed to manage the land, or to enforce the law. In any event, there was no public opinion to support action against timber thieves and trespassers—who were, more often than not, regarded as upstanding citizens. Referring to timber depredations in his 1870 official report, one high Washington official argued that the whole public domain should be disposed of : "The entire standing army of the United States could not enforce the regulations," he said. "The remedy is to sell the lands."

Timber was not the only resource plundered by the raiders. The soil that had been eons in the making was devastated in some regions as thoroughly as were the forests.

In 1852, Anthony Chabot, a California gold miner, ingeniously devised a canvas hose and nozzle that would wash banks of gold-bearing gravel into placer pits for processing. Chabot's labor-saving short cut caught on in the Mother Lode country above Sacramento, and by 1870 his crude hose had evolved into the "Little Giant," a huge nozzle that could tear up whole hillsides.

The result of the hydraulic mining was the massive movement of soil into the rivers that drained the Sierra Nevada. For every ounce of gold collected, tons of topsoil and gravel were washed into the river courses below. With the spring floods, clear streams became a chaos of debris, rocks, and silt; communities downstream were inundated with muck, and fertile bottomlands were blanketed with mud and gravel. The town of Marysville, along the Yuba River, was forced to build ever-larger levees that rose higher than the city's rooftops. In 1875, a big storm sent the Yuba over the levees and filled the city with silt.

The townsmen and farmers who suffered damage protested, but they got nowhere. Gold was California's first industry, and by the standards of the times, the hydraulic miners had as much right to slice off the hills as the farmers had to cultivate the valleys below. Minerals are not a renewable resource, and the land legacy of any mining operation is, necessarily, a pit, a shaft, or a hole. However, in their reckless effort to extract gold, the hydraulic miners asserted the right to damage other resources irreparably —and the homes of other citizens as well. In the 1870's, the right of each individual to do as he pleased was sacrosanct. The miners had enough friends in Sacramento to enable them to continue their onslaught on the mountainsides until the devastation forced the California legislature, in 1884, to outlaw hydraulic mining altogether.

Like the raids on lumber and gold, the raid on oil and natural gas was a strip-and-run business, with enormous incidental waste. When one field was depleted, the oilmen moved on to another. Where one well might have sufficed to drain an underground reservoir, there were often fifteen or twenty.

The big oil strikes began in the 1860's and, as with lumber, a succession of towns across America boasted, for a time, the title of "Oil Capital of the World"—Titusville, Pithole, Oil City, and Bradford in the Appalachian area; Tulsa and Oklahoma City in mid-continent; Kilgore in Texas; Los Angeles in California.

When oil came, it was usually in gushers that spewed the black liquid across the landscape with volcanic force. In the first big oil boom in Pennsylvania, just after the Civil War, gushers wasted oil at the rate of 3,000 barrels a day. But that record was soon broken and, in 1901, Spindletop in Texas flowed wild for nine days at the rate of 110,000 barrels a day before it was brought under control. The gushers went uncontrolled because early oilmen did not understand geology. Gushers caught fire, oil was allowed to evaporate in earthen dams, or to escape down creeks and gullies in an orgy of waste.

When early disputes arose as to ownership, the courts held that the oil belonged to anyone who could capture it. This Law of Capture put a premium on speed, and most of the time the big rewards went to whoever struck the underground treasure first. Consequently, operators who wanted to conserve supplies were forced to compete with greedy and shortsighted men.

Of all the waste associated with petroleum, perhaps none was so great as the waste of natural gas. The oil producer, not realizing that gas energy brought oil to the surface, allowed it to dissipate.

When it came to petroleum, the Myth of Superabundance reached an absurd climax. It was widely believed that oil was continuously being formed in the earth and was thus, like every other resource, supposedly inexhaustible.

A similar lack of land insight was responsible for the worst shortcoming of our stewardship, the appalling erosion of topsoil. Some of our farmers and stockmen knew what they wanted, but they knew too little about the delicate balances of nature. For example, deterioration of the Western grasslands by overgrazing was usually a slow process. It was hard for a rancher to notice each spring that the grass on his range grew a little thinner, and that, as the years passed, the invasion of weeds expanded. But as the overgrazing continued, scrub growth took over, the slopes began to wash, small gullies developed into large ones, and ranchers found themselves the proprietors of homemade badlands.

The farmers who tried to raise crops in the grassland country ran into similar trouble. Many regions west of the 100th meridian should never have been plowed at all: the familiar pattern of farming in the East was out of place in a region of little rain. In this country of half-steppe, half-desert, the soil was anchored to the land by grass. Once the plains were plowed, the dry, upturned soil had no protection against the driving winds. In most areas the soil might have been saved if the farmers had planted cover crops or had limited their plowing to the lands they intended to farm intensively. They failed to realize, however, that drought was a recurrent fact of life on the Great Plains, and when the dry years came, crops withered, the dust deepened, and whistling winds lifted thousands of tons of topsoil into the atmosphere.

The farmers' failure was a failure to grasp elementary earth facts. Like tobacco and cotton farmers of the South, they abused the land because they were ignorant of its laws of self-renewal. Aldo Leopold, who later looked with a scientist's eye on the prairie country, saw the subtle interrelations the settlers had missed:

> The black prairie was built by the prairie plants, a hundred distinctive species of grasses, herbs and shrubs; by the prairie fungi, insects, and bacteria; by the prairie mammals and birds, all interlocked in one humming community of cooperations and competitions, one biota. This biota, through ten thousand years of living and dying, burning and growing, preying and fleeing, freezing and thawing, built that dark and bloody ground we call prairie.

Defeated by dust storms of their own making, thousands of farm families packed up their belongings and retreated eastward. On the Great Plains, from the 1880's on, the tides of migration rolled back and forth with the weather.

After each rainy cycle of sod-busting, the dusty land waited for the wind. The settlers thus set the stage for the Dust Bowl of the 1930's—the most tragic land calamity ever to strike the North American continent.

In the long run overgrazing and overfarming proved as disastrous as overmining and overlogging. Yet the raids on resources were not limited to soil or gold or timber. They extended to biological resources as well. The Big Raid on wildlife began when the *voyageurs* and the colonists found that pelts would fetch a good price in European markets. It entered its final phase when the first major expedition of mountain men went up the Missouri to trap beaver. Within a decade, John Jacob Astor's American Fur Company and its competitors could ship pelts by the hundreds of thousands to Europe, where they were converted into men's high hats.

The trappers who were making a livelihood and the organizers like Astor who were making fortunes from beaver failed to realize that they were destroying a species. Owing to a happy accident, however, the beaver raid stopped short of extermination—the beaver hat went out of style by 1840 and the beaver was saved.

The shifting fashions were not so kind to another aquatic mammal, the fur seal of the North Pacific Ocean. In the late eighteenth century, seamen had often observed the northward migrations of enormous seal herds. In 1786, a Russian navigator, Gerasim Pribilof, gazed in unbelief at some volcanic islands in the Bering Sea where millions of seals blackened the shore lines. The enterprising Russian planted a colony on one of the lonely islands and piled his ship high with pelts. For eighty years Russian ships raided the fabulous rookeries of the Pribilof Islands. When the Czar agreed to sell Alaska to the United States for $7,200,-000 in 1866, the seal population, estimated originally at 5,000,000 had been cut in half.

The Americans of the Alaska Commercial Company, which received the United States franchise for the Pribilof furs, proceeded to outdo the Russians in slaughtering the animals. They launched a promotional sales campaign that soon brought competitors into the Bering Sea—seagoing hunters who shot the animals in the ocean during their migrations.

As the Pacific whaling industry declined, whalers turned to sealing, and by 1880 seal hunting on the high seas was a big business. Most of the animals which were shot were gravid, and many were not recovered; as a result, the waste was enormous.

The United States government showed less interest in the depletion of the herds than in the revenues paid by the sealers. In the first twenty years of its operation the Alaska Company took enough sealskins to repay the entire cost of the Alaska Purchase. By 1911, when the Fur Seal Treaty with Canada, Japan, and Russia was finally approved, only 3 per cent of the original seal population remained. If the treaty had been delayed another few years, the fur seal might well have become the first major marine victim of the Myth of Superabundance.

Of all wildlife species on the continent the most numerous was the passenger pigeon. At the beginning of the nineteenth century the number of these birds was estimated to be an incredible 5,000,000,000. Around 1810, Ornithologist Alexander Wilson reported sighting in Kentucky a single flock which was a mile wide and 240 miles long, containing, he guessed, more than 2,000,000,000 birds. It is likely that these prolific pigeons then constituted about a third of the entire bird population of the United States.

Succulent and easy to kill, they were shipped by the carload to city markets, and some farmers used them for hog feed. It was inevitable that these vast flocks would be depleted. They were an easy mark for the hunters, and the forest levelers were destroying their habitat. Ultimately, however, the passenger pigeons were not depleted—they were exterminated. At the end of the nineteenth century there was not one to be seen, and a few years later the last survivor of the species died in the Cincinnati zoo.

A good many other species of edible wild birds—robins, blackbirds, sparrows, thrushes—were in similar danger until the Audubon Societies and their allies took up cudgels for their protection. Some of the big birds whose bright plumage was much in demand for ladies' hats—the kingfishers, terns, eagles, pelicans, egrets, and herons—also came close to the abyss of extinction.

Many other native species barely survived the onslaught of the Big Raids. The sea otter was pursued to near extinction by rapacious seal hunters; the salmon runs were eliminated on nearly all New England rivers by mill dams and pollution; the whales were hunted relentlessly until commercial whale oil was replaced by petroleum.

That massive emblem of the American frontier, the buffalo—our largest and most valuable wildlife resource—was nearly extinguished in an unprecedented campaign of animal butchery. No one knows how many buffalo there were when Jefferson's Louisiana Purchase made their ranges part of our public domain. Informed guesses extend from 10,000,000 to 100,000,000, but whatever the exact number, they were the wildlife wonder of our continent.

The size of the buffalo herds was a source of awe to the plainsmen, and Colonel R. I. Dodge once wrote of a herd he saw in Arkansas in the early 70's: "From the top of Pawnee Rocks I could see from six to ten miles in almost every direction. This whole vast space was covered with buffalo, looking at a distance like a compact mass."

The big kill began when the Civil War ended. The army wanted the animals killed in order to starve out the Plains

Indians; the cattlemen wanted them killed to save forage for their own livestock; the railroad men wanted them killed to supply profitable freight in the form of hides; the market hunters wanted to kill them for their tongues and hides; and sportsmen came to kill them for trophies and for pleasure. For a few years buffalo hunting was the main, grisly business of the Plains country, and the more flamboyant of the market hunters—men like Bill Cody and Wild Bill Hickok—romanticized the last, gory raids and capitalized on their acclaim long after the great herds were gone.

The hunters roamed the bison ranges with specially designed rifles which could drop a buffalo on a far hillside. If the wind was right and if luck held, the hunters found a herd and picked off dozens of animals before the rest panicked. The westward push of the railroads was a boon to the hunters, and the organized slaughter began in earnest when the tracks reached the Plains. A good hunter could bring in more than 1,000 hides a year, but the venison—more succulent by far than the steak of a longhorn steer—was left for the wolves and vultures. An estimated 1,000,-000 buffalo were killed each year from 1872 to 1875. At Forth Worth, hides awaiting shipment were stacked in high rows over a quarter-mile long.

As the massacre continued, there was concern in some quarters that these shaggy symbols of the Wild West would be completely annihilated. But General Phil Sheridan boasted that the buffalo hunters were doing more to subjugate the Plains Indians than the army had been able to do in thirty years. In scorched-earth language the general advised the Texas legislature: "Let them kill, skin and sell until the buffalo is exterminated, as it is the only way to bring about lasting peace and allow civilization to advance." For the same reasons, in 1875, President Grant vetoed a buffalo-protection bill, the first measure ever passed by the Congress of the United States to protect a species of wildlife.

Within five years the Southern herds had been wiped out, and the buffalo hunters turned their attention northward. In the fall of 1883 a hunting party came upon a group of 10,000 of the animals near the Cannon Ball River in North Dakota. By November this herd had been finished off, and the hunters gathered up their hides and retired for the winter. The following spring the Indians and white hunters waited for the annual migration to begin, but they had done their work too well. Only a few hundred scattered survivors could be found. The raid on the buffalo was over.

Back as far as the 1830's, artist George Catlin had suggested that large areas should be set aside for the buffalo—and the Indians—as a great outdoor museum of natural history. If his advice had been followed, more than the buffalo might have been saved: it is ironical that the very grasslands that were the natural home of these magnificent beasts later became the scene of the overgrazing and overfarming that caused the dust bowl.

Entwined with the tragedy of the buffalo was the tragedy of human blindness that depicted the national mood. Only a handful of men spoke up to protest the slaughter, and, worse, the trustee of all the people, President Ulysses S. Grant, cast his lot with waste and butchery. In an era of superabundance and government inaction, public resources owned by everyone were, in practice, the responsibility of no one. By a fortunate accident some remnant herds in Yellowstone and in Canada were overlooked by the hunters. Their descendants that remain today on a few Western wildlife refuges are a vivid symbol of the most savage hour in the rape of American resources.

Most of the raids on wildlife, like those on other resources, were carried out in a devil-take-the-hindmost spirit under the deadly assumption that the supply was unending. No matter how many buffalo were shot, there would always be more. Bemused by the Myth of Superabundance, Americans ignored the elementary laws of nature. They realized too late that the Atlantic salmon would never come leaping up the rivers again, the sky would never again be filled with passenger pigeons, and the buffalo would no longer make men pause in awe at the thunder of their passing.

The brand of extreme individualism that necessarily characterized the frontier dominated our attitudes toward all resources during the nineteenth century. The intentions of those who launched the assault on our land were, however, as diverse as the American way of life itself.

Some who unwittingly diminished the productivity of our land were not raiders at all, but imprudent husbandmen like the early tobacco and cotton farmers who "wore out" two or three farms in a futile and halfhearted search for the secrets of soil fertility. The homesteaders and their sons who plowed up the Dust Bowl were honest husbandmen who never understood the Draconian laws of drought and the importance of grass cover in an arid land. The buffalo hunters, by contrast, realized full well that they were liquidating a resource, but considered themselves patriotic outriders of the civilized advance.

There were others, however, who found themselves caught up in a greedy game in which success went to those who cut the most corners and got to the most valuable resources first. The winners of these ruthless races were acquainted with hard work and sharp practice. They were men of pitiless initiative, with a total indifference to the needs of future generations. The most destructive of all the raiders were the hydraulic miners and the loggers who set fire to the large areas they had stripped: their wastelands and "barrens" stand today as monuments to a form of *laissez-faire* individualism that ignored sound husbandry.

This reckless era was also the time when Manifest Destiny came to full flower and we completed the settlement of the frontier. It had episodes of heroism and colonizing and empire building, but it should not be forgotten that it was also the period when we raided the Indians, raided the continent—and raided the future.

Nature, at times of her own choosing, would belatedly present her bills for waste and mismanagement to other generations. We would not acquire the kind of land wisdom we needed until we learned that nature's laws were paramount, that science and research held the keys to wise husbandry, and that government action was essential to save a permanent public estate of wildlife and water and forests and parklands.

In the era of the Big Raids, we desperately needed men who had their eyes on the horizon, men attuned to the wise ways of science and of politics. Such men began to emerge in the decades following the Civil War. They were not heeded at first, but their influence eventually grew into a tide that checked the raiders and began the saving of the American earth.

CHAPTER VI

The Beginning of Wisdom:

GEORGE PERKINS MARSH

Men now begin to realize what as wandering shepherds they had before dimly suspected, that man has a right to the use, not the abuse, of the products of nature; that consumption should everywhere compensate by increased production; and that it is a false economy to encroach upon a capital, the interest of which is sufficient for our lawful uses.

—GEORGE PERKINS MARSH, 1847.

By 1850 a sharp contrast in land attitudes in the United States had begun to develop. The old approach was typified by Marcus Whitman and the Mormons, who believed that the earth was the Lord's, and practiced the assiduous husbandry they preached; and the new, by the go-getters who knew full well that the earth belonged to the man who got there first, kept his claim stakes firmly fastened, and reaped while the reaping was good. Different groups, emigrating westward over the same trails, had diametric expectations of the earth. It was symbolic that Mormons with shovels were building irrigation canals and planting orchards in Utah while the forty-niners raced by en route to Sutter's Mill.

Settlers in many regions came to stay and put down

roots, but it was inevitable in a virgin land that raider attitudes would predominate. The continent had to be "conquered," and the new captains of capitalism, rushing westward to harvest its abundance, had what amounted to a crash program for national growth. These men were the doers and movers; under the creed of *laissez-faire*, their very success as creators of new technology and new wealth seemed to make long-range thinking on resources irrelevant.

However, even as early as Jefferson's presidency, solitary voices were raised in dissent. Edmund Ruffin, a perceptive Virginian, looked with dismay on the worn-out, gullied plantations of his state and inveighed against the ruinous practices of the tobacco farmers. Alarmed by erosion and land waste, Ruffin went up and down the Southern countryside preaching contour plowing, careful drainage by the use of furrows, soil rejuvenation by lime and fertilizers, and planned crop rotation. Sensing the approaching clouds of the Civil War, he warned the plantation owners of the "growing loss and eventual ruin of your country, and the humiliation of its people, if the long-existing system of exhausting culture is not abandoned." He advised his neighbors to survey the available remedies and to "Choose, and choose quickly!"

Another pre-Civil War dissenter was Eugene Hilgard, who might be called the father of modern agronomy in our country. His study of the dwindling yields of Mississippi farmlands in 1860 made it clear that the fertility of the soil was exhaustible. He, too, prescribed contour plowing and fertilizing to renew the soil and condemned careless farming:

Well might the Chickasaws and Choctaws question the moral right of the act by which their beautiful, parklike hunting grounds were turned over to another race. . . . Under their system, these lands would have lasted forever; under ours, in less than a century the State will be reduced to the . . . desolation of the once fertile Roman Campagna.

In the mid-1860's, as the Civil War drew to a close and the final western push was about to begin, we needed above all a geographer par excellence, who could expand on the insights of Ruffin and Hilgard, challenge the Myth of Superabundance, and provide an agenda for land rescue and land renewal.

It was our country's good fortune that such a land philosopher appeared. He was a Vermonter, and his name was George Perkins Marsh. The statement he offered his countrymen was a book, *Man and Nature*. Published first in 1864, it became a conservation classic.

Marsh had a mind as wide-ranging as Jefferson's and an interest in nature as broad as Thoreau's was deep. To him, the world of nature was second nature, and he studied its intricacies all of his life. He was, as he once wrote, "forest born" and on his father's Green Mountain farm he had at his doorstep a Walden of his own where he formed a fellowship with country things.

He once wrote that in his youth "the bubbling brook, the trees, the flowers, the wild animals were to me persons, not things." His favorite axiom was "Sight is a faculty, seeing is an art." He was intrigued by the interplay of natural forces and by man's role as an agent of change.

Given his head, young Marsh might have become a Dartmouth professor, but family pressure prevailed and at twenty-five he found himself a leading lawyer and businessman in Burlington. He was not satisfied, however, to be a modest main-street success. He restlessly read great books and pored over scientific papers, and his reading soon led him onto the main highroads of the mind. By the time he was thirty, he had mastered twenty languages. His omnivorous intellect ranged from Herodotus to Icelandic grammar, and he found time to translate Goethe, probe into Danish law, keep up with the latest developments in European silviculture, and pursue an ever-widening group of intellectual interests.

Science-minded and skeptical, the young New Englander rejected the easy idea that any resource could be exploited without concern for the future. Soil had a precarious purchase on the Green Mountain slopes. As a sheep raiser, lumber dealer, and farmer, Marsh saw the damage done to the valleys of Vermont by land misuse. Overgrazing and overcutting were eroding hillsides; rivers overflowed; wool-growing waned; and as the woods were logged out, Burlington, once a lumber boom town, was forced to import timber from Canada to keep its mills and factories from closing down. Men were working against themselves, and by 1840 much of the Green Mountain country was already an exhibit of improvident land management. Not even in the cotton and tobacco belt were soils exhausted faster or forests mangled more thoroughly than on the hillsides of Vermont.

As an eyewitness to the deterioration of his native ground, Marsh watched the mistakes of the past eroding the prosperity of the present. He had good reason to question a basic tenet of the Myth of Superabundance—the assumption that progress was inevitable. Progress was possible, he was convinced, only if men used wisdom in managing resources. He determined to look deeper into the economy of nature, and his keen interest in what we now call "ecology" stemmed from indigenous insights in the valleys of Vermont.

Marsh was already probing for answers to the problems of wise land use when he was persuaded to run for Con-

gress in 1842 as a high-tariff Whig. He was elected and
headed for Washington, excited at the prospect of an opportunity to rub shoulders with large-minded men.

The young Representative quickly formed fruitful contacts with many scientists and scholarly diplomats, but the
congressional colleague who did the most to crystallize his
thinking on resources was the pungent "Old President,"
John Quincy Adams, who had returned to public life as a
Congressman after his defeat by Andrew Jackson in the
election of 1828.

Adams' foresight qualifies him today as one of the un-
acknowledged prophets of the conservation movement. As
President, he rejected, in tones that would have warmed
the heart of Theodore Roosevelt, the "giveaway" approach
of the General Land Office, and he had nothing but contempt for speculators and Congressmen who, "with the
thirst of a tiger for blood," were attempting to seize public
lands. With hardly a pause to calculate the political risks,
he dusted off Jefferson's modest plan for "internal improvements" financed by federal funds, and proposed that it be
expanded and activated.

His ideas had a radical simplicity: land was the prime
source of national capital, and its resources should not be
dealt out to landgrabbers, but used to build roads and canals
and to promote science and education. Government, in
Adams' view, should be much more than a passive referee;
it should carry out positive plans to advance the common
good.

Adams' ideas anticipated the programs of both the
Roosevelts, and hence were unseasonable by nearly a century. In the 1820's the prevailing spirit of spoils, in both
politics and resources, doomed his land-use proposals from
the start, and helped defeat his bid, in 1828, for a second
term. Expedient politics might have kept him in office for

another four years, but his fierce honesty made him choose to be right rather than to continue as President. The climate of opinion was so unfavorable that his one significant stroke for conservation came to naught: in 1828 he had set aside a large live-oak forest in Florida as a source of supply for navy shipbuilders, but when Jackson took office he promptly canceled the order and threw the tract open to "the people."

Despite his lack of success as President, Adams made a lasting impression on Marsh, and the "Old President," in turn admired the discernment of his young colleague.

As an advocate of government support for science, Adams praised Marsh's plea in Congress for a research-oriented Smithsonian Institution as "one of the best speeches ever delivered in the House." He chortled at the prospect that the Philistine members would henceforth be forced to cope with what "many of them have never seen before, the spectacle of a living scholar." Nevertheless, the two New Englanders served in Washington at a time when the "least" government was the ideal, and during their years in the House together they had few opportunities to vote for resources or reform.

Marsh had his shortcomings as a statesman. He opposed our open-door immigration policies, and was convinced that the Western frontier "was unsuited to the genius of the people of the United States." He failed to anticipate that many of the immigrants, trained in Old World husbandry, would lead the way in the wise use of resources. His doubts about the frontier were not merely a result of New England provincialism, but stemmed from his instinct for putting first things first. The American people, he argued, should learn to use the land prudently before rushing pell-mell into unplanned exploitation of new territories. Marsh failed to gauge the force of Manifest Destiny as a national concept, although he saw its limitations with rare clarity.

Marsh was a better critic than planner, and his genius as a geographer lay in his ability to read the land record of the past. His appointment by President Taylor, in 1849, as Minister to Turkey opened a long diplomatic career that gave him a firsthand opportunity to study the geography of Europe and the Middle East as well as to form friendships with such diverse personalities as Matthew Arnold, the Robert Brownings, Ferdinand de Lesseps, Giuseppe Garibaldi, and Louis Kossuth.

Each of his journeys was a voyage of discovery, and, as self-appointed fieldman for the Smithsonian, he sent a steady stream of specimens to the Washington curators. He inspected much of the Nile Valley, took a camel caravan to the Holy Land, saw 'Aqaba and the ancient ruins of Petra, went mountaineering in the Alps, studied glaciers, scrutinized Vesuvius in eruption, traveled through the desolate Adriatic Karst, walked the eroded Tuscan hills, and saw the ravaged islands of Greece. Here were ancient lands that revealed the successes—and failures—of the "husbandry of hundreds of generations." For Marsh, every violated valley and hillside was pregnant with meaning.

It was a cosmopolitan, land-wise George Perkins Marsh who, in 1861, was appointed Minister to Italy by Abraham Lincoln. In a quiet retreat on the Italian Riviera he began to sort out the accumulated insights of half a century and record them in the book that would be his *magnum opus*.

It was Marsh's omni-competence, his wholeness as a man, that made *Man and Nature* a bench mark. Within his mind there was an incessant dialogue between a naturalist, a humanist, a historian, a geographer, and a practical politician; it was this versatility which gave him dominion over a wide range of human knowledge. *Man and Nature* was at once an exciting introduction to the inchoate science of ecology, a veritable encyclopedia of land facts, and the major American contribution to geography in the nineteenth century. The book gained an international reputation, and ten years after its publication in America a European reviewer remarked that *Man and Nature* had "come with the force of a revelation."

As Marsh worked, his huge study was piled high with books, periodicals, letters, and reports. The author's facility as a linguist gave him immediate access to the latest and best thought in western Europe. The bibliography he appended to *Man and Nature* included over two hundred items published in Finland, Holland, Denmark, Italy, Norway, Germany, France, Sweden, Spain, Switzerland, Belgium, England, Scotland and Austria, and ranging over a wide variety of subjects from soils to astronomy to philosophy.

Marsh succinctly announced the object of his inquiry in the preface: ". . . to point out the dangers of imprudence and the necessity of caution, in all operations which, on a large scale, interfere with the spontaneous arrangements of the organic and inorganic world. . . .

"The result of man's ignorant disregard of the laws of nature. . ." was the deterioration of the land, and Marsh painted the picture with a broad brush:

> The ravages committed by man subvert the relations and destroy the balance which nature had established. . . .; and she avenges herself upon the intruder by letting loose her destructive energies. . . . When the forest is gone, the great reservoir of moisture stored up in its vegetable mould is evaporated. . . . The well-wooded and humid hills are turned to ridges of dry rock, . . . and . . . the whole earth, unless rescued by human art from the physical degradation to which it tends, becomes an assemblage of bald mountains, of barren,

turfless hills, and of swampy and malarious plains. There are parts of Asia Minor, of Northern Africa, of Greece, and even of Alpine Europe, where the operation of causes set in action by man has brought the face of the earth to a desolation almost as complete as that of the moon. . . . The earth is fast becoming an unfit home for its noblest inhabitant, and another era of equal human crime and human improvidence . . . would reduce it to such a condition of impoverished productiveness, of shattered surface, of climatic excess, as to threaten the depravation, barbarism, and perhaps even extinction of the species.

To optimistic Americans these were extravagant words. Marsh was a Jeremiah prophesying doom, and it is understandable that men misled by the superabundance of a virgin continent would find his warnings farfetched.

Marsh, however, did not limit himself to lamentations over the errors of the past; most of *Man and Nature* was devoted to ecology; he inquired into the balance of nature and the interrelationships of plant and animal communities. He pointed out that every part of such a community, from microscopic organisms to earthworms to birds to trees to mammals, had its particular place in the web of life. Destroy any part of the web, Marsh warned, and the entire community might be disrupted. "Man is everywhere a disturbing agent," he wrote, "wherever he plants his foot, the harmonies of nature are turned to discords."

He deplored, for example, the "wanton sacrifice of millions of the smaller birds, which are of no real value as food, but which . . . render a most important service by battling, in our behalf, as well as in their own, against the countless legions of humming and of creeping things, with which the prolific powers of insect life would otherwise cover the earth."

On the other hand, the insects themselves, if their numbers were kept in balance, played an indispensable role:

"By a sort of House-that-Jack-Built, the destruction of the mosquito, that feeds the trout that preys on the May fly that destroys the eggs that hatch the salmon that pampers the epicure" could result in a severe depletion of fish.

In the same way, nature's balance could be disastrously upset by the larger works of man. Marsh pointed out the ramifications of local climate and geography resulting from the draining of lakes and marshes, and the alteration of riverways and coast lines. Fascinated as he was with engineering projects—canals, dams, and wells—his interest extended to their side effects on water tables, wildlife habitats, vegetative cycles, and the micro-climates. His scientific skepticism led him to suspect that all single-minded "improvements" by man resulted in unforeseen harm, and he reasoned that "we are never justified in assuming a force to be insignificant because its measure is unknown, or even because no physical effect can now be traced to it."

In his book, Marsh described the usefulness of the minute "animiculae" in the soil; discussed the influence of the forest on soil and weather; analyzed the flow of sap; evaluated the destructive action of floods; inquired into the origin of sand and of mountain lakes; wondered about the effect of drifting sand on the Suez Canal and the possibility of damming the wadies of Arabia—and described and interpreted a hundred assorted phenomena. To him the lesson in each case was plain: to disturb the balance of nature without calculating the consequences was to invite disaster.

Marsh thought the restoration of many of the disturbed natural harmonies was possible: man should co-operate with nature to correct the harm he had unwittingly done.

. . . He is to become a co-worker with nature in the reconstruction of the damaged fabric which the negligence or the wantonness of former lodgers has rendered untenantable. He must aid her in reclothing the mountain slopes with forests and vegetable mould, thereby restoring the fountains which she provided to water them; in checking the devastating fury of torrents, and bringing back the surface drainage to its primitive narrow channels.

In Marsh's time, the agricultural, geologic, and atmospheric sciences were in their infancy. Although many empirical pronouncements had been made, there was little theoretical understanding by which these could be related to each other. Hence, his primary scientific contributions were the questions he asked. In his preface he avowed: "In these pages it is my aim to stimulate, not to satisfy, curiosity, and it is no part of my object to save my readers the labor of observation or of thought."

He was well aware that many of the answers would await the accumulation of basic scientific data, and he urged that there be intensive meteorological, topographical, and hydrological surveys throughout the world. As the instrument that both developed and replenished resources, science could produce "new triumphs of mind over matter." Once men learned how to utilize science, rivers could be harnessed, water conserved, forests renewed. By well-planned projects, men could reclaim and revitalize the earth.

Marsh did more than any of his countrymen to deflate the illusion that the resources of the United States were somehow self-renewing and inexhaustible. "It is certain" he wrote, "that a desolation, like that which has overwhelmed many once beautiful and fertile regions of Europe, awaits an important part of the territory of the United States . . . unless prompt measures are taken to check the action of destructive causes already in operation."

As long as Americans were obsessed by the Myth of Superabundance they could rationalize the waste of resources, and it would be impossible to persuade them to adopt measures of prudent management. *Man and Nature* did not destroy the myth, but it punched holes in it and for the first time let the acid of doubt seep in.

The book gave strong support to the advocates of a forestry program for the United States. The woodlands were the most endangered resource, and Marsh devoted nearly half of his text to this problem. His experience in Europe and the Middle East had crystallized his Vermont impressions, and he asserted that reckless burning and indiscriminate cutting were "the most destructive among the many causes of the physical deterioration of the earth."

Pointing to floods and erosion, he stressed the vital function of forests as watersheds. These spongy conduits of our hydrologic system were nature's chief conservers of water and soil, and it was plain that the forest would save more water if science were given the task of prescribing cutting and replanting practices. Marsh was convinced from the results of German and French silviculture that lumbering would be more efficient if trees were grown artificially on farms, and if cutting were restricted to the harvest of mature timber. The intensively managed forests he saw in Europe produced, over the long run, a greater volume of wood, with less waste, than did clear cutting of natural growth.

Marsh urged that American landowners reforest the woodlands in recognition of the "duties which this age owes" to the next. He was fifty years ahead of his countrymen in this call for a national program of experimental forestry. If research on biological improvement of trees was "the work of centuries, so much the more reason for beginning now," Marsh advised. Unfortunately, however, Americans of his day were far too busy cutting trees to care about improving them.

As a Congressman, Marsh had hoped that education and enlightened self-interest would induce the lumbermen to alter their wasteful practices, but as the years passed and the devastation widened, he concluded that national leadership was needed to save the remaining forests.

Marsh's interest in hydrology led him beyond forest lands to regions of the earth where water is scarce, and in 1874 he prepared, at the request of the Commissioner of Agriculture, a paper on the feasibility of irrigation in the arid West. Irrigation projects would be possible, he wrote, if undertaken on a river-basin scale, and based on thorough hydrological surveys. He foresaw the dangers of private monopolies of water supplies, and his report concluded that all Western irrigation should be "under Government supervision, from Government sources of supply."

Many of Marsh's questions still cannot be fully answered today: What are the long-range results of man's modification of the environment? When men clear a forest in order to make space for agriculture, how does this clearing affect the climate, the rate of erosion of soil, and the populations of birds and other wild animals? How much of the water that formerly seeped into the ground is now lost by evaporation or by rapid runoff?

Marsh's call for greater research efforts in these fields is still valid. He was most significant, however, as the framer of a new land ethic. Without restraint derived from a sense of respect for the land, he was convinced the new machines would only hasten the exploitation of the raiders. Science, guided by a new land conscience, should have an opportunity to give primacy to the needs of coming generations; the success of our stewardship would be measured by the extent to which the land was redeemed and enriched for those to follow. Man was part of the cycle of nature, and the fall of a sparrow or the felling of a tree should be studied in the context of the total environment. These concepts were Marsh's most valuable contributions, and in time they became part of a saving American creed.

Here then, is a prescient book, with the scientific and moral foundations of a new land policy approach. George Perkins Marsh has been called the "fountainhead of the conservation movement," and with the perspective of a full century we can perceive that his words—and those of his fellow dissenters—were the beginning of land wisdom in this country.

CHAPTER VII

The Beginning of Action:

CARL SCHURZ AND
JOHN WESLEY POWELL

Powell was one of those powerfully original and prophetic minds which, like certain streams in a limestone country, sink out of sight for a time to reappear farther on. . . . He tried to shape legal and political and social institutions so that they would accord with the necessities of the West. He tried to conserve the West's natural wealth so that it could play to the full its potential part in the future of the United States. He tried to dissipate illusions about the West, to sweep mirage away. He was a great man and a prophet. Long ago he accomplished great things and now we are beginning to understand him . . . even out West.

—WALLACE STEGNER
Beyond the Hundredth Meridian

MARSH'S APPEAL from Italy for the scientific management of resources reached a small but influential group of scientists and sensitive men in the United States. It would take painful political scandal and scandalous land waste to make most Americans ponder the conservation ideas Marsh espoused.

The last gun fired at Appomattox was, in effect, the starter's gun in an intensified race for resources. The war served as a spur to technology, and aided by new machines, the raiders set off in hot pursuit of wealth. In 1865, Northerners who were interested in resources and who could put together new pools of corporate venture capital sensed that success was in the air and some of them found it in quantities beyond all imagining. The result was the "great barbecue" of resources described by the historian, Vernon Parrington. The Myth of Superabundance helped encourage reckless exploitation, and the folk belief which portrayed the Great American Desert as the "Garden of the World" added allure to the unsettled areas of the West.

A central circumstance in the postwar years that eliminated many of the old restraints was the turn of the political wheel that brought the raider spirit to the highest councils of government. From the day General Grant took office in 1869 the main business of government was the business of abetting the raid on resources. In willing partnership with compliant Congressmen who perfected the pork barrel and shared in the raider profits of Credit Mobilier, Grant became our only full-time, full-fledged failure as President. He was the grand chef of the Great Barbecue, and his failure was largely failure to conserve the national estate—to grasp the moral lapse involved in extending spoilsmanship from politics to the public domain.

President Grant was the people's trustee (whether he chose to acknowledge it or not) and his appearance in public with the notorious Jay Gould—whose greed and audacity tempted him to corner the gold reserves of the United States—made speculation respectable and set the stage for the grossest giveaway in our history. The most lavish gifts of national lands to railroad promoters were made under Grant, and the consummate lobbyist of the period was certainly Jay Cooke, who garnered a grant of land larger than Missouri for his Northern Pacific Railroad. Theoretically, the purpose of these grants was to encourage the westward extension of the nation's railroads by supplying bonuses to their builders. But twenty years after Cooke's road was built, the aggressive James J. Hill built a parallel and highly profitable railroad without any kind of subsidy from the public purse—and revealed the prodigal dimensions of the Cooke giveaway.

High-grade ores, timber, wheat, and oil had markets overnight once the new machines were put to work and the railroads fanned out across the land. Instead of following the "grow-slow-and-learn-husbandry-first" policy suggested by Marsh, the resourceful entrepreneurs relied on a grow-fast policy—and to quicken the pace of progress some of them sent agents abroad to encourage immigration. The hungry immigrants dug the mine shafts, built the railroads, and felled the forests. Their cheap labor provided much of the true capital which laid the foundations of industrial America.

For nearly a generation the moguls who jostled at the barbecue table dominated American life, and as their power expanded, it posed the threat that interlocking resource monopolies would remake our political system into an instrument of the economic royalists.

One counterforce to this trend was the monumental Homestead Act signed by President Lincoln in 1862, which terminated a long quarrel over land policy. This land-reform legislation embodied the idea that each man should have an opportunity to own a 160-acre parcel of the American earth if he had the ambition to earn it by elementary acts of husbandry.

As a plan for national development, homesteading had its shortcomings, but it established a new kind of freeholder, and these new sons of free soil became, in time, a salutary counterbalance to the raiders. The homesteaders were a plains-and-prairie version of Jefferson's yeomen, and the seer of Monticello would have applauded when they organized the Granger and Populist protests that helped check the onrush of monopoly capitalism in the 1890's.

The lot of the homesteaders was hard, and their lonely lives are a stretch of sadness in our history. The harshness and heroism of homesteading life is a saga in itself: these men were empire builders of another sort, and their story —an essential part of the American epic—has been recounted in literature by Willa Cather, Hamlin Garland, Ole Rölvaag, and other sons and daughters of homesteaders.

Though successive scandals rocked Grant's administration, he and his cronies had to depart before the protest against waste could find an effective voice. When Rutherford B. Hayes entered the White House in the spring of 1877, he appointed, as Secretary of the Interior, Carl Schurz, a crusading Senator from Wisconsin, who had the Marsh approach to resources. As an immigrant refugee from Germany after the unsuccessful Revolution of 1848, Schurz had brought with him to the New World a knowledge of the forestry practices of his native land.

Schurz's name in our history is linked with the movement for civil-service reform, but for a brief period, at a crucial juncture, he was a land reformer as well. In particular, he sounded a rallying cry and excited the first hope that the remaining forests might be saved.

Schurz had the energy to gather his own facts, and the courage to act on them. He had a lively sense of stewardship and a highly developed concept of the public trust. His gifts as an orator and journalist enabled him to lay before the American people a powerful case against the predators of the public forests.

His first act as Secretary was to initiate an intensive study of forest depredations, and his first report, in 1877, singled out lumbermen who were "not merely stealing trees, but whole forests." Schurz pointed out that the cut-and-run policies of the timber barons would, in due course, destroy the self-renewing character of the woodlands. However, he soon discovered that his fieldmen in the General Land Office, who were supposed to be looking after the forests, were spoils appointees inclined to wink at trespass and timber theft.

Schurz sent a few enforcement teams into the field, made some trespassers pay for their board feet, and exposed one lumberman who had brazenly established a milltown on public land. He quickly proposed such stern remedies as the establishment of a system of federal forest reserves, the initiation of reforestation practices, charges to the users of national resources, and stiff penalties for the willful setting of forest fires. In addition, he urged that Congress empower the President to appoint a commission to study the "terribly instructive" laws and practices of other countries.

His recommendations drew a contemptuous response from timber-state Congressmen. Speaker James G. Blaine of Maine called the Schurz program outrageous and un-American: it would, he charged, introduce "Prussian methods" into a democratic country, and oppress honest enterprisers who were the backbone of the republic. To put an end to the argument, Blaine and his colleagues chopped off Interior's meager appropriation for timber inspectors. The people's forests were unwatched and unmanaged, but Schurz, with fresh support from the newly organized American Forestry Association, hobbled ahead with a limited campaign of protest and reform.

"Deaf was Congress, and deaf the people seemed to be," he said later, but his foursquare pronouncements even now have the ring and fervor of Teddy Roosevelt's best conservation stump speeches. Schurz's 1877 report is also a landmark, for his dissent helped set in motion forces that encouraged others to act.

The following year there came across his desk a conservation white paper more significant than his own. Prepared by Major John Wesley Powell of the Interior's Powell Survey, it was entitled "A Report on the Lands of the Arid Region of the United States." Major Powell sought to strike blows of his own while a conservation-oriented secretary was seated as his superior, and was very probably encouraged to do so by Schurz.

One of Powell's contemporaries described him as a man "electric with energy and ideas." After losing an arm at Shiloh, he had returned from the war to a restless two years of college teaching in southern Illinois. In the summer of 1867, and again in 1868, he had led an oddly assorted scientific expedition to the Colorado Rockies.

Excited by the wide open country, he spent a year organizing a daredevil team of river runners, who plunged down the wild chasm of the Colorado River in the summer of 1869. This, for Powell, was far more than an adventure down the wildest river on the continent. It was a mission of science, and his search for specimens, his geologic inspections, his map work and scientific observations were carried out in the face of physical hazards that cowed some members of his expedition.

Before he was through, he had accumulated a storehouse of land wisdom about the West. A friend once said he knew more of the live Indian than any other man, but knowledge of Indians was only a piece of him: Thoreau himself would have admired Powell's sensibilities, and he had the curiosity and physical daring of Jedediah Smith. Moreover, he studied the problems of land and people with the broad gaze of a Jefferson, and his interest in the climate and resources of the vast provinces of nature, his understanding of water and watersheds and the role of man as a geologic force in a fragile landscape made him the intellectual blood brother of a man he never met—George Perkins Marsh.

For nearly ten years after his river expedition Powell studied the ecology of the plains and high plateaus as intensively as Charles Darwin studied the life of animals. The cycles of rivers and rainfall, the village life of the Utah Mormons and the land-use ideas of the Indians and Spanish-American settlers enabled him to grasp the essentials of order in the West. He saw that water was the critical resource in a region of little rain, and his 1878 report was a broad conservation plan for the settlement of the arid country—a plan which included Jeffersonian political and social institutions adapted to the special conditions of the West.

Because of his knowledge of the essentials of Western living, Powell was an instinctive enemy of the myths and the mythmakers. In a harsh and inhospitable land men would court disaster, unless they came to terms with nature.

The report Powell presented to Schurz and the Congress was, in essence, a land-use plan for the western part of the United States. His conclusions were starkly simple. He pointed out that the land west of the 98th meridian was arid; that it received, except in a few areas of the Pacific Coast and in the mountains, less than twenty inches of annual rainfall—not enough to sustain an economy based on traditional patterns of agriculture; and that many of the homesteaders beginning to move out into the arid lands were therefore doomed to fail.

He concluded that dry-land farming would depend on

irrigation; and that the best opportunity for irrigation would lie in the development of the large streams, where the construction of large dams and canals would require federal leadership and federal financing. Since there was not enough water to go around, it had to be shared equitably. The irrigable land would be only a small fraction of the whole, but the flow of smaller streams should be conserved to further the development of holdover storage reservoirs downstream. Prudence dictated that the best reservoir sites should be identified and reserved at the outset, and it was obvious that new water-rights laws and new forms of co-operation would be needed if any democratic plan of settlement was to be carried out.

In the West, Powell pointed out, land by itself was almost worthless; it could be made permanently valuable only by water. Water rights were, therefore, more valuable than land titles, and should be tied to each tract by law. Already there were ominous signs: huge cattle corporations were staking out dukedoms in the public domain by controlling all the water of large areas, and some settlers were already quarreling over water rights. For the arid lands there was no workable water law. The riparian rights of English law had been designed for areas where water was abundant, but beyond the 100th meridian the old laws and old rules did not fit. In wet countries, water could be used at streamside for a mill without being *used up;* after use it was returned to the stream. In the arid West, when water was taken out of a stream and spread on a field, it was consumed and the downstream flow was diminished. Each irrigable tract, then, should have its own appurtenant water right.

The central unity of the West was the unity of little water, and Powell saw that the whole region would come to grief unless land policies and political and social institutions were shaped in accordance with its peculiarities of climate. Settlement would have to work *with* nature, Indian fashion, as the Spanish colonizers and Mormon settlers had done. Powell's report embraced the related problems of water, erosion, floods and soil conservation, and even foreshadowed the coming problem of hydroelectric power. Major Powell believed in the small farmers and freeholders, and saw the need to protect them not only from speculators, but also from natural conditions they did not understand and could not combat. False estimates of fact and climate, he knew, would lead only to disaster, and the time to plan was before the patterns of settlement were fixed. The plateau country west of the Rockies, as well as much of the short grass plains, was as yet largely unsettled, and Powell saw that there was still an opportunity to plan the use of water and land if the best experience of the Spanish and the Mormons was put to use.

Powell was more than a generalizer; his basic conclusions called for drastic reforms in the use of land and water, and he attached to his report sample bills which spelled out his ideas for legislation. For one thing, the Homestead Act needed to be revised. Where irrigation water was plentiful, most farmers found that a tract of 80 acres was as large an area as they could cope with. In the high country of the West, 160 acres were either too much or too little—too much for an irrigation farmer, and far too little for a cattle ranch. Faced with this fact, the government should revise the land laws and adapt them to the necessities of the arid regions.

In irrigable valleys, under the Powell plan, nine or more settlers would join together to form irrigation districts and apply to the federal government for a survey. Surveys should proceed not only by the rectangular system of ranges, townships, and sections which had been sacrosanct since Jefferson invented it, but according to watersheds and drainage basins. Water access was absolutely essential, and plots should be shaped by the terrain and not be over 80 acres in size. With water, 80 acres of Western lands constituted an economic unit that one family could manage. Unless watersheds were protected and water rights ran with the land, as Powell's plan provided, the water monopolists would cripple the well-rounded settlement of the West.

The pasture lands of the Plains were a special problem. In the aftermath of the Civil War, enormous herds of cattle moved in search of the stirrup-high bluestem, the succulent switch grass, the Indian grass, and the waving fields of grama and buffalo grass. The cattlemen's "rush" was a rapid one, and within the short period of fifteen years well-heeled cattle companies (many involving English and Scottish investors) invaded the empire of grass. Walter Prescott Webb described the arrangement of the open range this way:

> As yet no ranch man owned any land or grass; he merely owned the cattle and the camps. He did possess what was recognized by his neighbors (but not by law) as range rights. This meant a right to the water which he had appropriated and to the surrounding range. Where water was scarce the control of it in any region gave control of all the land around it, for water was the *sine qua non* of the cattle country. In the range country "divides" became of much importance, marking the boundary between the ranch men of one stream valley and those of another. Up and down the same stream the problem was not quite so simple, but the ranch men were careful to recognize that possession of water gave a man rights on the range. Moreover, it was not good form to try to crowd too much.

It was only a matter of time until overgrazing damaged millions of acres of public rangeland, and even in Texas, where nearly all lands remained in private ownership, the grassland displacement and erosion was severe. The grass rush, while it lasted, was a big bonanza, but the men of the cattle kingdom overplayed their hand and nature soon caught up with them. On the Northern Plains, the cattle industry (including a North Dakota cattleman named Theodore Roosevelt) sustained overwhelming losses in the harsh winter of 1886, and in the Southwest drought decimated most herds in 1893. The old land lessons were learned anew as ranges deteriorated, drought cycles returned, and serious soil erosion began.

As the buffalo were eliminated and the Plains Indians were brought under control, the cattle herds grew in size each year. With overcrowding, the inevitable conflicts came. For a while the Code of the West—the unwritten law of the cowmen—kept the peace, but in the end ugly conflicts arose on every side.

On the public-domain part of the grasslands the rule was always first come first served. There were in the grass country even fewer federal land managers than in the forests, and in the absence of either landlord or referee, there began an era of range wars between cattlemen and sheepmen, between cattlemen and cattlemen, and between stockmen and homesteaders. The grass suffered, livestock suffered, the stockmen themselves suffered, and large areas of land were impaired.

The cattle invasion of the Plains country was already under way when Powell submitted his report, but he hoped the mistakes could be remedied before irreparable damage was done. Under his plan, settlers could organize themselves into pasturage districts, and government surveys would carve out livestock ranches covering four whole sections and a water right. (All Western experience had shown that at least four sections—2,560 acres—would be needed to support a family through stock-raising.) Ranch residences should be grouped for community purposes, and the range should be a jointly managed common pasture. Powell observed that grasslands needed careful management, and he foresaw that overgrazing would turn the plains into desert.

It was as simple and orderly as that. Powell's "General Plan" would have incorporated the best experience of the cattlemen and the best co-operative experience of the Mormon and New Mexican irrigators and ranchmen. It would have prevented the worst dust bowls of the Great Plains, and it would have produced an equitable sharing of water and land among the oncoming settlers.

But like Carl Schurz's forest plan, Powell's reform proposals got a cold reception. For one thing, his plan flew in the face of the holy Homestead Act—and worse, he spoke about shortages and the climactic shortcomings of the West at a time when promoters were still saying that "rain would follow the plow." His report used bear language in a bull market, and most of the Western leaders would have none of it.

Powell did achieve one lasting result: he struck a blow for science. At his urging, the United States Geological Survey was created to find facts about the earth so that sensible resource planning could begin. Science finally had a tax-supported home in government, and basic research began as an enterprise of all the people. The many mansions of science in our Federal establishment today rest on the foundation stones laid by such men as Powell, Clarence King, and the pioneers who, in 1873, organized the American Association for the Advancement of Science. To Powell, science was the one discipline that might renew and enlarge all resources in the long run, and the more he saw of men who were blindly conquering the land, the more he became convinced of the importance of science as a tool of national progress.

His report on the arid regions was too sweeping, settlement was too sparse, and experience too meager, for Powell to win his argument in the late 70's. However, his book was dusted off when the toll of human disaster taken by the droughts of the 80's and 90's drove his points home. He got a second chance when Congress passed crisis legislation in 1888 and he was put in charge of an Irrigation Survey to select reservoir sites, determine irrigation project areas, and carry out part of his General Plan.

Powell thought it might take six or seven years, and cost six or seven million dollars, to get the answers and lay out reservoir sites and canal lines needed for a sound irrigation program. In 1889 he carried his message to the constitutional conventions of North Dakota and Montana, and he told the newly forming Western states that their land values had to be measured in acre-feet: irrigation, he said, was a necessity; county lines should follow drainage divides, and each river valley should become a political unit where men could work together rather than at cross purposes. He reiterated his idea that each landowner should have a corresponding water right—a concept which later became the "Wyoming doctrine" and fixed the water-rights pattern for most of the states of the West.

But time and impatient politicians conspired against him. With one more year's work, he might have been ready to activate plans for some valleys, but members of the "irrigation clique" in the Congress could not wait, and in 1890 they killed the Irrigation Survey.

The General Plan was dead. Four years later Major Powell retired from the Geological Survey to spend the rest of his life writing anthropology and philosophy in his office in the Smithsonian. The arid region continued to develop by trial and error, and Powell's defeat meant that from 1890 until the dark skies of the "dirty thirties" finally forced a change, the rangelands would deteriorate under the pounding of too many hoofs.

But Powell's irrigation ideas were to be vindicated. The Reclamation Act, passed in 1902 (the year he died), was not everything he had fought for, but it was a beginning —and the face of the West today approximates, in most valleys, the face Powell envisioned.

Schurz and Powell were, in a real sense, our first field generals in the crusade for land reform and land preservation. As reformers, they lost the battle with their own generation. As land prophets, they won, and their insights have become basic concepts of the conservation movement. We have yet to pay them due honor, but the historian, Bernard De Voto, once passed this judgment on the arid lands report of John Wesley Powell:

In the whole range of American experience from Jamestown on there is no book more prophetic.

CHAPTER VIII

The Woodlands:

PINCHOT AND
THE FORESTERS

The American Colossus was fiercely intent on appropriating and exploiting the riches of the richest of all continents—grasping with both hands, reaping where he had not sown, wasting what he thought would last forever. New railroads were opening new territory. The exploiters were pushing farther and farther into the wilderness. The man who could get his hands on the biggest slice of natural resources was the best citizen. Wealth and virtue were supposed to trot in double harness.

—GIFFORD PINCHOT
Breaking New Ground

THE timber barons and their congressional allies, who had run roughshod over Carl Schurz in the late 1870's, continued to ride high both in the West and in Washington. In the 70's and 80's permissive timber laws were passed, which opened new avenues to those intent on taking trees by trespass and fraud. The penalty for stealing a horse in some parts of the West was often death, as Gifford

97

Pinchot wryly noted, but the stealing of public trees cast not a "shadow on the reputation of the thief."

The unlimited frontier freedom, which had swept aside the old barriers of caste and class, quickly erected new ones of wealth and gave political power to the new rich. Hand-picked Congressmen dedicated to the Great Giveaway were the chief spokesmen in Washington for the landgrabbers, and for a time the domineering men who ran the company towns turned such states as Maine and Wisconsin and Oregon into company states.

Gifford Pinchot saw the climax of the Big Raids in the 1890's, and he described the scene with both candor and color:

> Out in the Great Open Spaces where Men were Men the domination of concentrated wealth over mere human beings was something to make you shudder. I saw it and fought it, and I know. . . . Big money was King in the Great Open Spaces, and no mistake. . . . The powers and principalities which controlled the politics and the people of the West began to emerge from the general landscape. Principalities like the Homestake Mine in the Black Hills, the Anaconda Mine in the Rockies, Marcus Daly's feudal overlordship of the Bitterroot Valley, and Miller and Lux's vast holdings of flocks and herds and control of grazing lands on the Pacific slope— these and others showed their hands or their teeth. So did powers like the Northern and Southern Pacific and the Great Northern Railroads, the irrigation interests of California, and the great cattle and sheep stock growers' association.

In the forests, as on the ranges and in the mines, it was every man for himself, and it would take a generation of protest, and a Rough Rider President, to slow down the onslaught and put the get-rich-quick capitalists on the defensive. The nineteenth-century lumber tycoons, to give them full credit, housed a growing nation, cleared land,

and hastened the pace of westward expansion. However, in the process, they set world records for waste, and their prodigal prosperity consumed the stored "capital" of nature —which, by right, belonged to other generations.

By our standards they were spendthrift, but they played under spendthrift rules, and a mill operator who stopped to save soil, or protect second growth, or reduce the danger of fire would have gone bankrupt in a hurry. Lumbering was our largest manufacturing business during most of the nineteenth century, and the stripped hillsides and blackened woods left by its reckless "rush" did more than anything else to awaken us to the fallacy of the Myth of Superabundance.

Before the forest raids were finished, about half of the cutover woodlands had gone into farms and the other half was in sorry second growth or had been logged and burned into barrenness. By 1920 only one-fifth of our primeval forest lands remained uncut. This was one of the most wasteful chapters written by the raiders.

Most of the time, the American pioneer was too busy subjugating the land to be interested in husbandry. Even the first isolated acts of preservation were dictated by necessity, and not by wisdom about the woods: the early colonial governors who blazed selected trees with a "broad arrow" did so for the sole purpose of reserving superior shipmasts for the British fleet; and the common woodlots maintained by some New England towns were attempts to save enough nearby trees for local home building and native crafts.

Although from independence onward, disposal was our land policy, in the early 1870's small countercurrents were set in motion. The need for systematic preservation first became apparent in the Eastern states where the raids had begun and where the consequences of forest waste and land damage were already visible. The first government agency to reverse the trend was the New York Forest Commission, which, in 1872, halted the sale of state forest land.

The following year, some members of the American Association for the Advancement of Science opened a campaign to save the forests, and another strong voice was added two years later when the American Forestry Association came into being. These pioneer conservation organizations carried on a constant campaign for a new approach to national stewardship.

Meanwhile, in the hinterland, the tree raid was still moving toward its climax, but a few Eastern magazines and newspapers were beginning to raise doubts about the race for resources, and the need for national policies was increasingly apparent. The abortive efforts of Schurz inspired the introduction of initial forestry bills into the Congress. Senator Edmunds of Vermont, George Perkins, Marsh's nephew, twice secured Senate approval of a bill to establish a forest reservation "at the headwaters of the Missouri River." But the House failed to pass either measure, and new giveaway bills had an easy time in Congress and opened up further opportunities for the landgrabbers.

The forest savers made slow headway during the decade of the 80's, but their persistent campaign finally elicited support from President Harrison and his Secretary of the Interior, John W. Noble. In the closing hours of an 1891 congressional session—largely through Noble's adroit use of a veto threat, these sixty words were inserted, as a "rider" in a public lands bill:

Section 24. That the President of the United States may, from time to time, set apart and reserve, in any state or territory having public lands wholly or in part covered with timber or undergrowth, whether of commercial value or not, as public reservations, and the President shall by public proclamation, declare the establishment of such reservations and the limits thereof.

This terse proviso did not, on its face, seem to threaten the Great Giveaway. Noble's amendment was ungrammatical and apparently innocuous, and its potential did not penetrate the minds of the adjourning members. Committees had not considered it; its adoption contravened the rules of both Houses; it was not debated; and had its scope been spelled out, the representatives of the forest states of the West would have opposed it almost to a man.

As a piece of legislation, it was a fluke: one of the most far-reaching conservation decisions ever made was ironically consummated in half-hidden haste. At one stroke it gave Presidents the power to set aside or "withdraw" certain forests where all logging would be forbidden, and to control the disposal of the remaining forest lands.

Within a month after passage of the act, President Harrison withdrew 13,000,000 acres and set up 15 "reserves" under Noble's Department. As a people, we now had some land that was off limits for the forest raiders—at least on paper. However, no one had a management plan, for in 1891 there was only one professional forester, Bernhard Fernhow, a German emigrant of 1876, who was a friend of Schurz and tried to implement some of his ideas.

Without rules for use, this reserving of certain forests, as some Westerners rightly pointed out, was a "lock-up" of land. A showdown was inevitable, and it came in 1897 when Grover Cleveland—then a lame-duck President—signed a proclamation that set aside new forest areas double the size of the Harrison reserves.

Outraged Senators demanded impeachment of the President, and bills were passed to reverse Cleveland's proclamation. Congress would have repealed the 1891 act then and there, but a presidential veto and adjournment saved the day. The stage was thus set for Gifford Pinchot, who strode onto the Washington scene in 1898 with a plan and a program for the systematic management of American forests.

Pinchot, then thirty-eight, was wellborn, well-educated, and well prepared to inaugurate practical forestry in America. After Yale he studied silviculture in France with a master teacher, Sir Dietrich Brandis, the founder of forestry in British India, and verified Marsh's earlier observations on European forest practices. On his return to the United States, young Gifford made long field trips into the back country of the South and the West, and later was hired to manage the Vanderbilt estate in the Great Smoky Mountains.

Appointed as the Department of Agriculture's Chief Forester in 1898, he headed an information service that had no timberland to manage. But he soon set about to remedy the situation. The man and the job were made for each other : "GP" had the clear eye of a scientist, a naturalist's love of woods and open spaces, the moral fervor of an evangelist, and a politician's intuition. Before he was through he would need all of these talents, for if he was to succeed, the flow of American thought about natural resources would have to be reversed.

He had the lean look of a Moses and a similar sense of mission. His task, he was convinced, was to stop the practice of forest destruction in the United States, and to inaugurate the practice of forestry. He knew the conditions on America's watersheds, and once he decided on a solution he was as single-minded as the timber raiders themselves. He had long argued that the idea of "reserves" was wrong : the ax should be regulated, not stopped in mid-air. He believed any forestry plan would work only if the people who lived nearby would support it. This concept became the heart of Pinchot's plan. His instincts told him that the Western men in Congress could not long tolerate huge, permanent land withdrawals that threatened the development of basic industries. Pinchot devised a use-with-a-long-view plan, and his compromise approach halted those who wanted to revoke the Cleveland proclamation.

The big break for Pinchot, and for forestry, came when Theodore Roosevelt assumed the presidency on the death of McKinley in 1901. The two men were already acquainted. As Governor of New York, Roosevelt had called Pinchot in for consultations about the resource problems of his state. In his first State-of-the-Union message the new President declared that "the forest and water problems are perhaps the most vital internal questions of the United States." (This part of his message had been written by Pinchot and F. H. Newell, one of Powell's former aides who was then the chief advocate of federal leadership in Western water development.) It included a call for a federal reclamation program and a sound use plan for the "reserves."

During Theodore Roosevelt's first term, Pinchot worked overtime to sell his idea that forests could be "saved" and used simultaneously. He was still a forester without forests, and his objective was to persuade the President and the Congress to give him a corps of trained forest rangers and make him the warden of all federal timberlands.

A tireless and persuasive talker, GP began to cultivate support in and out of Washington. He knew as much about the growth of public opinion as he did about the growth of trees, and each major step he took, or proposed to the President, was calculated to catch a fresh breeze of public support. The study commission, and the national conference were tools he perfected to further his cause. Once TR was elected in his own right in 1904, Pinchot arranged a Washington meeting of the American Forestry Congress to coincide with the President's long-awaited recommendation on forest management and a bill was swiftly passed in 1905, transferring the reserves from the Department of the Interior to the Department of Agriculture, and they were soon designated National Forests.

By any standards, Gifford Pinchot was a magnificent

bureaucrat. In his time the Forest Service was the most exciting organization in Washington. It was more a family
than a bureau. In the field, around campfires, and in his
home GP discussed the next moves and gave his associates
the feeling that they served on the general staff in a national crusade. A natural leader, he chose his men well,
gave them authority, aroused an *esprit de corps* and sent
them forth to save the forests. The rule book he wrote for
his men was filled with crisp, common-sense guidelines.
Typical Pinchot maxims were: "The public good comes
first," and "Local questions will be decided by local officers
on local grounds." In a matter of months his new "forest
rangers" were winning over the West.

Pinchot was a fighter who gave the raiders blow for
blow. At a Denver conference one day, a cattle industry
spokesman who charged a ranger with outrageous conduct
was put to silence when Pinchot called his bluff: "If what
you say is true, give me the man's name, and he will be
fired today!" He had no use for the monopolists and the
free-loaders on the public domain, and they quickly found
out that the public interest, for him, was more than a
phrase. Grazers and loggers were assessed user fees, and
for the first time grass and trees were managed in such a
way that they could be replenished on a sustained-yield
basis. Law and order came late to the public lands, but
Pinchot's rangers prevailed because they were firm and
fair and spoke for the future.

Despite the initial successes of the Forest Service, the
die-hard opponents of federal forests watched uneasily as
the young President signed a series of land-conserving
proclamations. They were convinced a line had to be
drawn, and in 1907 a group of Western Senators sought to
turn the tables on Pinchot and Roosevelt by attaching to
an appropriation bill a rider that repealed the Forest Reser-

vation Act of 1891 in the timber-rich States of Oregon, Washington, Idaho, Montana, and Wyoming.

It was the end of the session, and the President was forced to sign the bill, but the alert Pinchot was ready with a counterstroke. Ten days later the floor of the President's office was strewn with maps and surveys as the final decisions were made, and TR, with inimitable gusto, signed proclamations creating 16,000,000 acres of new forests in these five states. Later that day, his conservation goal accomplished, he inked the bill that would have tied his hands. All told, under the Act of 1891, Presidents Harrison, Cleveland, and Roosevelt set aside 132,000,000 acres which today comprise the greater part of the magnificent national forests of the West.

Pinchot and TR would lose some battles, but they had already won the most important engagements in the conservation war. Once Pinchot had saved his trees, he turned his attention to other abuses of the raiders. One February day in 1907, while Pinchot was riding his horse through Rock Creek Park in Washington, the conservation cause took on a new aspect:

> Suddenly the idea flashed through my head that there was a unity in this complication—that the relation of one resource to another was not the end of the story. Here were no longer a lot of different, independent and often antagonistic questions, each on its own separate little island, as we have been in the habit of thinking. In place of them, here was one single question with many parts. Seen in this new light, all these separate questions fitted into and made up the one great central problem of the use of the earth for the good of man.

On that day in 1907, three hundred years had elapsed since the founding of the first colony at Jamestown—three hundred years of exploitation and misuse of the continent's resources. On that day Thomas Jefferson had been in his grave for eighty-one years, George Perkins Marsh had been in his for twenty-five, and John Wesley Powell had been dead for five. On that day nearly all of the buffalo were gone, most of the forests had been leveled, and millions of acres of soil were eroding away. But a countermovement was at last well under way, and Pinchot had defined it: "the use of the earth for the good of man."

What was needed now was a word, a name—to sum up the concept. A conversation with forester Overton Price brought up the fact that government forests in India were called "conservancies." Pinchot and Price liked the ring of the word, and thus a concept that had originated in the seminal thinking of such men as Thoreau and Marsh now had an expressive name—conservation.

Pinchot was one of the great teachers of his time: he taught frugality when waste was the accepted creed; he turned his back on the race for riches and sought the higher goal of public service; and when money power was king in parts of our land, he aroused in the people a sense of their own power.

It was the Chief Forester who framed most of the ideas which became Theodore Roosevelt's conservation program. The influential White House Conference on Conservation in 1908, the farsighted Inland Waterways Commission study, and the landmark report of the International Conservation Commission in 1909 were Pinchot-planned and Pinchot-executed projects. He was the composer, and also played first violin while the redoubtable Roosevelt conducted.

In all our history, Pinchot is the only bureau chief who was first counselor to a President on high policy. TR's letter to Pinchot the day the President left the White House was a memorable commendation: "As long as I live

I shall feel for you a mixture of respect and admiration and an affectionate regard. I am a better man for having known you. . . . I owe you a particular debt of obligation for a very large part of the achievement of this Administration."

Pinchot stayed on with Taft, but their convictions differed, and the tandem harness did not fit. At best, Taft was lukewarm toward TR's policies, and the contrast of personal styles was, to use Pinchot's words, "as though a sharp sword was succeeded by a roll of paper, legal size." Within a year the Chief Forester deliberately picked a fight with Taft over the actions of his Interior Secretary, Richard A. Ballinger. The still-disputed Ballinger-Pinchot controversy led to Pinchot's stormy departure, and also generated the friction which ignited the first fires of the Bull Moose movement.

The conservation movement was a river of many tributaries, and if GP was not, as he liked to believe in his later years, its fountainhead, he was nevertheless one of its vital sources. He was key man of a key decade, and his leadership was crucial in persuading the American people to turn from flagrant waste of resources to programs of wise stewardship.

Once launched, the forestry movement quickly gathered momentum. In 1911, Eastern Congressmen and conservationists passed the Weeks Act, and the policy of complete disposal had come full circle: from Maine to Florida, the government began to buy back cut-over tracts for inclusion in a national-forest system for the East. Public lands were indispensable after all, as Easterners belatedly discovered.

The preaching of Pinchot and his men—and the public opinion they stirred up—began to penetrate the lumber industry itself. A few leaders began to wonder if Pinchot's sustained-yield idea was not worth a try, and "tree farming" under private land management had a hopeful beginning.

Pinchot had shortcomings, and one of them led him to an unhappy confrontation. Over the years the extraction aspects of land conservation became a fetish with him. He always had a blind spot to wildlife and wilderness values: to him, untrammeled wilderness was a form of waste. It took the later leadership of foresters like Robert Marshall and Aldo Leopold to establish a system of wilderness reserves in the high country of the national forests. The nascent national park concept also left Pinchot cold.

The classic encounter that embroiled him and other dedicated conservationists involved the first head-on collision between the Pinchot idea of conservation-for-use and the park concept of scenic preservation. The controversy was known by an improbable name: Hetch Hetchy. Pinchot's chief opponent, a friend who knew even more about the American out-of-doors than he, was an improbable person —John Muir.

CHAPTER IX

Wild and Park Lands:

JOHN MUIR

There is an eagle in me and a mockingbird . . . and the eagle flies among the Rocky Mountains of my dreams and fights among the Sierra crags of what I want . . . and the mockingbird warbles in the early forenoon before the dew is gone, warbles in the underbrush of my Chattanoogas of hope, gushes over the blue Ozark foothills of my wishes —and I got the eagle and the mockingbird from the wilderness.

—CARL SANDBURG
"Wilderness"

A VIOLENT GALE was roaring through the forests of the Sierra Nevada one cold, clear morning in December of 1874 as John Muir wandered through the woods, "enjoying the passionate music of the storm." It occurred to him that a tree was the best place to catch the full force of a windstorm, so he chose a 100-foot Douglas fir, scrambled to the upper branches, and swayed there for hours "like a bobolink on a reed."

"Never before," he wrote later, "did I enjoy so noble an exhilaration of motion . . . My eye roved over the piny hills and dales as over fields of waving grain, and felt the light running in ripples and broad swelling undulations across the valleys from ridge to ridge, as the shining foliage was stirred by corresponding waves of air."

The episode was typical. Muir was determined to stretch his senses to their limits. He invariably sought the most intimate relation to natural forces—to the winds, the storms, the rivers and forests and wild animals—whether it required climbing to the top of a wind-bent fir or sleeping out in a mountain blizzard to feel the snowflakes on his face. He developed his inner ear to catch the nuances of nature, and this led him to evolve land-preservation concepts that still have a unique purity and simplicity.

John Muir's lifelong education in what he called the "University of the Wilderness" began when he was a boy in Scotland and continued in youthful wanderings in the back regions of the Great Lakes. It reached a broader phase in late 1867 when he took a 1,000-mile saunter through Kentucky, Tennessee, Georgia, and Florida. His odyssey continued to Cuba, through the Isthmus of Panama, and on to San Francisco.

He was thirty when he first saw the Golden Gate and set eyes on the summits of the Sierra Nevada. It was the spring of 1868 and he knew at once he had found his homeland. He described with absolute rapture the great range and the Central Valley, brilliant with wild flowers: "And from the eastern boundary of the vast golden flower bed rose the mighty Sierra, miles in height, and so gloriously colored, and so radiant, it seemed not clothed with light, but wholly composed of it, like the wall of some celestial city."

In his first exploratory walk into this mountain wilderness, he viewed a steep-walled canyon which became for him a personal sanctuary—the stupendous valley of the Yosemite. From the first, Muir sought as many contacts with the wilderness as one man could absorb. To him, the true wilderness experience was far more than mere exposure to nature; it began with heightened sensibilities and ended in exactness of observation. His senses were alive to the most subtle variations of color, line, texture, aroma, and sound. In a description of Yosemite Falls, for example, he wrote that the plummeting water "has far the richest, as well as the most powerful, voice of all the falls of the valley, its tones varying from sharp hiss and rustle of the wind in the glossy leaves of the live-oaks and the soft, sifting, hushing tones of the pines, to the loudest rush and roar of storm winds and thunder among the crags of the summit peaks."

For Muir it was not enough merely to observe the fine sculpturings of the Sierra peaks, the lake basins carved in solid rock, the polished granite surfaces of canyon walls; he wanted to find the roots of things and he viewed the natural world with the eye of a scientist. His theory that Yosemite Valley had been carved by Ice Age glaciers, although ridiculed at the time by leading geologists (who believed that the valley's floor had dropped down in a cataclysm), was later confirmed. In describing the panorama visible from the top of Mount Ritter, Muir visualized the vanished glaciers that had carved the peaks, canyons, and valleys below:

Standing here in the deep, brooding silence all the wilderness seems motionless, as if the work of creation were done. But in the midst of this outer steadfastness we know there is incessant motion and change. Ever and anon, avalanches are falling from yonder peaks. These cliff-bound glaciers, seemingly wedged and immovable, are flowing like water and grinding the rocks beneath them. The lakes are lapping their granite shores and wearing them away, and every one of these rills and young rivers is fretting the air into music, and carrying the mountains to the plains.

Muir saw that men were eyewitnesses to creation if only they opened their senses to it. Each journey into the wilderness was for him a trip to a fresh wonderland. It was also an experience of self-knowledge and self-fulfillment. He felt the same reverence for the land—the sense of wholeness and oneness—that had been experienced by the Indians and the early naturalists. In the wilderness, he wrote, "life seems neither long nor short, and we take no more heed to save time or make haste than do the trees and stars. This is true freedom, a good practical sort of immortality."

Early in his mountain career Muir came to a conclusion that decisively affected his own future and to some degree the future of his country: wilderness freedom, like political freedom, was perennially in danger and could be maintained only by eternal vigilance. It was necessary, he became convinced, permanently to preserve large tracts of choice lands in public ownership.

Although many years earlier such naturalists as George Catlin, Emerson, and Thoreau had vaguely recognized the need to preserve some of our finest landscapes, the first specific steps toward doing so had been taken only a few years before John Muir arrived in California by a young landscape architect named Frederick Law Olmsted, the designer of New York's Central Park. Impressed by the grandeur of Yosemite Valley, Olmsted and others persuaded Congress to pass a bill to preserve it "for public use, resort and recreation." The measure was signed by President Lincoln in 1864. Yosemite Valley, ceded to the State of California and administered along lines suggested by Olmsted, was the first scenic reserve created by federal action, and the event is a landmark in the history of conservation.

A few years later members of an expedition exploring the Yellowstone region in Wyoming were so overwhelmed by the beauty of the geysers, canyons, waterfalls, lakes, and forests that they began to discuss ways and means of saving a few superlative parts of primitive America for all time. If these men had shared the raider mentality of their day, they might have staked out a commercial bonanza for themselves. They could, quite lawfully, have filed home-stead or mining claims in the nearest land office and ex-ploited key tracts of this masterpiece of nature for private profit.

But to some of these explorers monopoly of such scenery was unthinkable, and the idea of a permanent public re-serve was discussed over a campfire one night in 1870 at the confluence of the Firehole and Gibbon rivers. One of the explorers, Judge Cornelius Hedges, later wrote a news-paper article on the subject. The idea generated support and it was only two years later that President Grant signed a little-debated and little-understood Yellowstone Park bill, providing that more than 2,000,000 acres—a region larger than Rhode Island and Delaware combined—be "dedicated and set apart as a public park or pleasuring ground for the benefit and enjoyment of the people. . . ."

Because Wyoming was still a territory, Yellowstone was placed in the custody of the Secretary of the Interior and became our first national park. The concern of a few people for the rights of future generations made the difference, and this factor of foresight would mean the success of most future park proposals.

The remoteness of the Yellowstone region had protected it from the inroads of the loggers and cattlemen, but Muir's plan to create a comparable nationally owned park in the mountain area surrounding Yosemite Valley was another story. The users, already on the scene, had come to look on the Yosemite high country as their private pre-serve. Muir had seen the first encroachments on his sanc-

tuary early in his mountain career when he came upon de-
nuded "gardens and meadows" on the Merced River above
Yosemite Valley. The sheepmen and their "hoofed locusts"
were responsible, and Muir had written in anger: "The
money-changers were in the temple."

On trip after trip he had found evidence of the destructive results of overgrazing. ". . . The grass is eaten close
and trodden until it resembles a corral. . . . Nine-tenths
of the whole surface of the Sierra has been swept by the
scourge. It demands legislative interference."

Muir's search for a remedy to protect the wild lands carried him beyond Emerson and Thoreau and the earlier
naturalists to take up the fight for the land. Protest was
not enough; men had to act; and from that point on, John
Muir became a sort of Senator-at-large for the American
out-of-doors.

Sheep were not the only agents of destruction in the
Sierra. In exploring the groves of giant sequoia, Muir
found sawmills going at full capacity. Lumbermen were
cutting magnificent trees thousands of years old, blasting
them with gunpowder into manageable size, wasting half
the timber and setting fire to what was left. The sounds
of lumber-mill saws, "booming and moaning like bad
ghosts, destroying many a fine tree," still rang in his ears
when, in February, 1876, he wrote a newspaper article,
"God's First Temples—How Shall We Preserve Our Forests?", which was an appeal for the protection of the
forests.

Like George Perkins Marsh, he described the ultimate
result of the misuse of upland watersheds. The soil and
water of these mountain areas contained "the roots of all
the life of the valleys." In destroying the balance of nature
in the mountains, man was cutting his own lifeline.
"Whether our loose-limbed government is really able or

willing to do anything," he concluded, ". . . remains to be seen."

He did not wait to see, however. Although Muir shunned the public attention his writings would bring, he saw his course as clear. He had filled his pen in the wilderness; now he must direct it to write those thoughts that came to him out of the wondrous silence of a high mountain meadow, or flowed into him from the shiny granite as he rested and watched the clouds through pine boughs overhead. He found Eastern publishers eager to print not only his wilderness stories, but also his pleas for legislation to save the common wealth.

The tone of his writings ranged from lyrical to vitriolic. His sense of indignation was expressed in his opinions of the loggers who were felling the sequoias:

> . . . Through all the wonderful, eventful centuries since Christ's time—and long before that—God has cared for these trees, saved them from drought, disease, avalanches, and a thousand straining, leveling tempests and floods; but he cannot save them from fools—only Uncle Sam can do that.

Not satisfied with merely putting the burden on Uncle Sam, Muir, a city-hater, came down out of the mountains time after time to do battle for his wild lands. He failed in his first assaults on the Myth of Superabundance, but bills he inspired were introduced in the '80's to save the finest groves of sequoias as federally owned parks, and to enlarge the Yosemite reservation.

It took a series of articles by Muir, published by his ally, Robert Underwood Johnson of *Century* magazine, to arouse the country to the need for the preservation of the entire Yosemite region around the relatively small state-managed park in Yosemite Valley. Interior Secretary

Noble, one of Muir's readers, took up the cudgels with President Harrison and the Congress, and a special bill was passed, in October of 1890, creating a "forest reservation" of more than a million acres. At about the same time, other Muir plans were partially fulfilled when the Sequoia and General Grant reserves were created to save some of the finest remaining groves of big trees. Although the legislation did not specifically refer to these three areas as "national parks," the term had previously been applied to Yellowstone, and Secretary Noble adopted it in naming them.

Encouraged by this success, Muir set out to form a private organization of mountaineers and conservationists to carry on his fight for the wilderness. Out of this effort came the Sierra Club, a crusading organization ". . . to explore, enjoy and protect the nation's scenic resources. . . ." Second only to John Muir in marshaling the organization's forces was his young colleague, William E. Colby, who joined the club two years after its founding, was its secretary for forty-four years, and a leader of the organization for more than half a century.

With the formation of the Sierra Club, Muir's career entered a new phase: the writer-naturalist became an organizer and a politician. The Sierra Club's first big fight, under Muir's leadership, was a counterattack against lumbermen and stockmen who wanted a stronghold in the Yosemite country. A House-passed bill would have cut the park area in half, but the Muir men mounted a counteroffensive, and the bill was tabled by the Senate. The Sierra Club had proved that the California vigilante idea would work for conservation if enough determined men were aroused.

But there was more to come. Surrounded by Yosemite National Park, Muir's beloved Yosemite Valley itself was still in the hands of the state and badly mismanaged by profit-seeking concessionaires. Meadows were plowed, the forest floor was grazed bare, trees were felled, and Mirror Lake was dammed for irrigation. The floor of the incomparable valley was becoming a barnyard. In 1895 an outraged Muir and the Sierra Club opened a decade-long campaign for national management which could resist untoward pressures of local interests.

The turning point of the fight came when Muir and President Theodore Roosevelt crossed paths in 1903. The two outdoorsmen camped under the sequoias of the Mariposa Grove, rode horseback on the long trail to Glacier Point overlook, and talked late into the night around the fire. For once the ebullient Roosevelt met his match in conversation. Like all others who encountered Muir, Roosevelt was spellbound by the eloquence and enthusiasm of this bearded zealot who preached a mountain gospel with John the Baptist fervor. The mountaineer did nearly all the talking and the President listened, fascinated, while Muir denounced the damage being done in the Sierra by the loggers and the stockmen and expounded at length about nature and wilderness.

One night the two men rolled up in their blankets and went to sleep on the ground. Next morning they found themselves covered with four inches of snow, and when they rode down to the valley, the President rejoined his party, shouting, "This has been the grandest day of my life!"

Muir had argued strongly for recession of Yosemite Valley to the federal government, and the President left Yosemite a convinced "recessionist." But Roosevelt's approval was not enough. It was also necessary to convince the California legislature. The Sierra Club carried the campaign to Sacramento and won a hard-fought victory early in 1905. Writing to his friend Johnson, Muir described his adventures in politics:

> I am now an experienced lobbyist; my political education is complete. Have attended Legislature, made speeches, explained, exhorted, persuaded every mother's son of the legislators, newspaper reporters, and everybody else who would listen to me.

Recession of Yosemite Valley was now possible if Congress approved, and Muir recruited a new ally, the railroad tycoon E. H. Harriman. With Harriman and Roosevelt on their side, Muir and his associates were able to swing the necessary votes, and Yosemite Valley finally became part of Yosemite National Park.

During the first years of the twentieth century, the burgeoning conservation movement had established itself as a force to be reckoned with on the American scene. Pinchot and his men were forcing the commercial interests to use forests wisely, while the Muir men were bent on barring commercial activity altogether from some of the finest remaining primeval landscapes. Pinchot and Muir were fast friends in 1896 at the time they worked together with the National Forestry Commission. When the commission reached Arizona's Grand Canyon, the two left the party. Pinchot in later years described the occasion:

> While the others drove through the woods to a "scenic point" and back again, with John Muir I spent an unforgettable day on the rim of the prodigious chasm, letting it soak in. . . . When we came across a tarantula he wouldn't let me kill it. He said it had as much right there as we did.

The rest of the commission bedded down in a hotel, but Muir and Pinchot decided to sleep out on the rim of the canyon in freezing weather. "Muir was a storyteller in a million," Pinchot wrote. "We made our beds of cedar boughs in a thick stand that kept the wind away, and there he talked till midnight. It was such an evening as I have never had before or since."

That night on the rim of the Grand Canyon was almost the last time the two men were on speaking terms. A clash between them, which began soon afterward, was perhaps the most dramatic confrontation in the history of the conservation movement. It was, in a way, inevitable, since each was headstrong, opinionated, and on fire with a sense of mission.

They had fought on the same side in the first stages of the fight against the raiders and against waste and mismanagement of the national estate. Muir, in the prime of life when Pinchot had been a student at Yale, was a stalwart battler to save the forests, and his eloquent pen gave fresh courage to those who carried on the lonely struggle for forest protection in the 1880's. Both men were for federal reserves, for government action, and for scientific programs of planning and management.

It was understandable that Pinchot, trained in forestry, would place his major emphasis on silviculture, and on the development of a sustained-yield harvesting program for the forests. If he failed to comprehend the need for parkland preservation, it is explainable by the fact that he was compelled to concentrate on the rules and educational work necessary to establish order in the forests. If he emphasized product values and considered the esthetic values merely incidental this, too, was understandable.

Unavoidably, Pinchot's philosophy of conservation for use collided with Muir's conviction that the best parts of the woodlands and wilderness should be preserved inviolate as sanctuaries of the human spirit.

The conflict came, in the first instance, over sheep-grazing practices in the mountains of the West. Pinchot later admitted that "overgrazing by sheep does destroy the forest. . . . John Muir called them hoofed locusts, and he was right." But Pinchot was an adept politican and felt compelled to accommodate his plans to user-group demands whenever his long-range goals were not violated. He wrote: ". . . We were faced with this simple choice: Shut out all grazing and lose the Forest Reserves, or let stock in under control and save the Reserves for the Nation." Pinchot chose compromise, but Muir was unbending, and sheep-grazing, controlled or not, was anathema to him.

On the one hand, Pinchot looked on the public lands as a workshop to be managed for many purposes under a plan of balanced use. Muir accorded a place in the resource picture to livestock and hydroelectric power, but he gave first priority to preserving the finest landscapes of the public domain as temples unspoiled and intact. Drawing a line between the workshop and the temple was, and still is today, the most sensitive assignment for conservation planners.

Their sheep argument intensified, but the clashing concepts of these two giants finally came to a historic showdown on Muir's home ground, Yosemite National Park. Within the park, on the Tuolumne River some twenty miles north of Yosemite Valley itself, was another glacier-carved valley called Hetch Hetchy. "I have always called it the 'Tuolumne Yosemite,' " John Muir wrote, "for it is a wonderfully exact counterpart of the Merced Yosemite, not only in its sublime rocks and waterfalls but in the gardens, groves and meadows of its flowery park-like floor."

In the "Tuolumne Yosemite," officials of the growing city of San Francisco, searching for a new source of water and power found an "ideal" location for a dam site. They filed a claim on the valley in 1901 and the issue was joined, for the proposed dam and reservoir were squarely within the boundaries of the National Park.

The wide waters of the reservoir would obliterate a sublime valley for all time, but the city needed hydro power and an assured water supply and the resource-development conservationists found themselves in a fierce contest with the conservers of the parks.

The Muir men were convinced that the integrity of the whole national-park system was at stake. If Yosemite National Park could be invaded and the Hetch Hetchy valley inundated, no other park would be safe from the dam builders and advocates of water-power development. Obviously, the parklands would be a prime future target as it would be cheaper to build dams in public parks than to buy up private property elsewhere.

The fight continued for more than a decade and involved many of our national leaders. The reservoir and public-power advocates finally played their trump card—a report of an advisory board of army engineers, which pointed out that, although there were several other possible sources of water for San Francisco, Hetch Hetchy would be the cheapest to build and would generate the most electric power. The report defined the issue squarely in dollars-and-cents terms. John Muir replied with indignant eloquence:

> These temple destroyers, devotees of ravaging commercialism, seem to have a perfect contempt for Nature, and, instead of lifting their eyes to the God of the Mountains, lift them to the Almighty Dollar.
>
> Dam Hetch Hetchy! As well dam for watertanks the people's cathedrals and churches, for no holier temple has ever been consecrated by the heart of man.

The emotional argument raged from San Francisco to Washington as "nature lover" and "bird watcher" became opprobrious epithets. There was no middle ground for compromise, as Muir's friend, Congressman William Kent of San Francisco, sadly found out. Kent had purchased and given the Muir Woods grove of redwood trees to the nation as a tribute to the man he later defeated in the Hetch Hetchy fight. Muir's central contention was that other dam sites were available but were not given consideration by the Congress or by his opponents. A second-best dam site would have saved water and scenery, too, but that was not to be.

In December of 1913 a bill was passed and the Hetch Hetchy dam was authorized. To Muir, the lesson was plain: no wilderness anywhere, even in the national parks, could be kept unspoiled unless the believers in preservation of scenic masterworks learned to organize and marshal new strength. Sadly, John Muir, seventy-five and exhausted from the grueling battle, wrote: "They will see what I meant in time. . . ." Sensing that the end was near, he feverishly worked fifteen hours a day to finish a book of his Alaskan travels. In December of 1914 he died, the manuscript at his bedside.

The valley of Hetch Hetchy was a superior scenic resource of the North American continent. It was flooded out, but those who had fought a losing fight for the principles of park preservation served notice on the country that its outdoor temples would be defended with blood and bone.

Muir's career ended on a note of failure, but the Hetch Hetchy episode must be seen in larger perspective. During John Muir's lifetime, and to a large degree because of his leadership, the national-park idea became part of the conservation constellation. He played a vital role in establishing six of our superb national parks—Sequoia, Yosemite,

Mount Rainier, Crater Lake, Glacier, and Mesa Verde—
and a dozen parklike national monuments, including two
that eventually became national parks, Grand Canyon and
Olympic.

Among those who were inspired by Muir was Stephen
Tyng Mather, an energetic Chicago businessman. Recovering, in 1904, from a nervous breakdown, Mather had
turned to the mountains for rejuvenation, and had become
interested in the Yosemite recession fight. He joined the
Sierra Club, took part in the organization's annual mountain outings and first met Muir on a camping trip in the
summer of 1912. Like Pinchot and Roosevelt and all the
rest, Mather was caught up by Muir's eloquence and enthusiasm.

Two years later, indignant over land depredations in
Sequoia and Yosemite and appalled at the sight of cattle
grazing inside the parks, Mather wrote an irate letter to
Secretary of Interior Franklin K. Lane, and was promptly
invited to join his staff as the directing head of the national
parks. Mather accepted his offer in 1915, but at the outset,
he had very little to work with in the way of staff or funds.
From the beginning, the national parks had been an administrative stepchild of the Interior Department. Each
park superintendent was responsible directly to the Secretary of the Interior, who had little time to co-ordinate the
management of the parklands. For many years Muir, and
such colleagues as landscape architect Frederick Law Olmsted, Jr., and J. Horace McFarland of the American Civic
Association, had urged the formation of a special agency
to administer and protect the parks, and their urging
finally resulted in the enactment of the National Park
Service Act in 1916.

Self-made millionaire, philanthropist, mountain climber,
and promoter by nature, Mather, like Pinchot, an inspirer

of men, was an ideal father for the National Park Service. He used his business acumen and powers of persuasion to curb grazing privileges, protect migratory birds, conserve historical landmarks and enlarge the park system itself. He instituted an educational and interpretive program to make the parks meaningful to visitors, and when Park Service appropriations were not available, he used his personal checkbook to accomplish urgent ends.

Mather thought national parks should be spacious areas of superior scenery to be preserved forever for the highest forms of outdoor recreation. He persuaded both the Department of the Interior and the Congress to accept this definition, and slowly the country at large caught on to the essentials of our National Parks.

A dilemma that plagued Mather and his successors was inherent in the national-park concept. The act of 1916 instructed the Interior Department "to conserve the scenery . . . and the wildlife . . . in such a manner and by such means as will leave them unimpaired for the enjoyment of future generations." As the most popular parks attracted thousands of visitors, the problem of "impairment" became a grave one, and each year the National Park Service had to resolve its use-but-don't-spoil dilemma. Mather realized that the only solution was to build up a tradition of high standards through a corps of highly trained, dedicated personnel, who would make sound judgments with their eyes on the needs of other generations. As a result the National Park Service today exemplifies one of the highest traditions of public service.

When Mather took office in 1915, there were fourteen maladministered national parks. By the time failing health forced him to resign in 1929, there were twenty-one superb units under one creative plan of management. The American national-park idea is today a conservation ideal in all parts of the world—the most enduring tribute to Muir and to Stephen Tyng Mather.

More and more Americans see, as Muir did, that in this increasingly commercial civilization there must be natural sanctuaries where commercialism is barred, where factories, subdivisions, billboards, power plants, dams and all forms of economic use are completely and permanently prohibited, where every man may enjoy the spiritual exhilaration of the wilderness. Americans have belatedy begun to prize the values of their wild lands and parklands, and each year more of them see the significance of John Muir's good counsel:

> Climb the mountains and get their good tidings. Nature's peace will flow into you as sunshine flows into trees. The winds will blow their own freshness into you and the storms their energy, while cares will drop off like autumn leaves.

CHAPTER X

Men Must Act:
The Roosevelts and Politics

*In this stage of the world's history, to be fearless, to be just
and to be efficient are the three great requirements of national life. National efficiency is the result of natural resources well handled, of freedom of opportunity for every
man, and of the inherent capacity, trained ability, knowledge, and will—collectively and individually—to use that
opportunity.*

—THEODORE ROOSEVELT
Message to Congress, January 22, 1909

IT WAS no accident that Powell, Pinchot, and Muir were
all eventually pitchforked into practical politics. From the
80's onward events forced those who were determined to
save forests and parklands and wildlife to take their places
on the conservation battle line in Washington and at the
state capitals.

Thoreau, in an era of abundance, had looked down his
nose at government and stood aside as many things he
prized were "civilized off the face of the earth." But after
his Western counterpart, John Muir, had set the precedent,
no one who believed in saving the land and its resources
could in good conscience be aloof from the politics of conservation.

126

It was hard going in the 80's and the 90's for the early conservationists. Their few legislative victories, such as the 1891 Forest Reservation Act, were the result of accident or indifference, for the Big Raiders were ground-floor lobbyists who quickly learned ways to turn wealth into political power. To them, any proposed legal restraint on an enterprise was un-American, and they had a hundred years of history to prove that they were right.

It was, in the end, the reaction against this ultraindividualism that gave the conservation movement its first foothold. The activities of the Robber Barons—termed by Theodore Roosevelt the "tyranny of mere wealth"—ran counter to the democratic idea.

TR witnessed the events that led to the showdown, and his description of the state of affairs is still trenchant:

. . . During the century that had elapsed since Jefferson became President the need had been exactly reversed. There had been in our country a riot of individualistic materialism, under which complete freedom for the individual . . . turned out in practice to mean perfect freedom for the strong to wrong the weak. . . . In no other country in the world had such enormous fortunes been gained. In no other country in the world was such power held by the men who had gained these fortunes; and these men almost always worked through, and by means of, the giant corporations which they controlled. The power of the mighty industrial overlords of the country had increased with giant strides, while the methods of controlling them, or checking abuses by them on the part of the people, through the government, remained archaic and therefore practically impotent. . . ."

The overlords and monopolists and landgrabbers had already overreached themselves by the time TR arrived in Washington as Vice President, and their actions precipitated a crisis of politics that coincided with the land crisis.

This showdown made it plain that the only power that could assert the national interest was the power of the federal government.

The man the American people re-elected in 1900, William McKinley, was an ideal presiding patriarch of the *status quo*. Passive and tractable, he had a proper respect for wealth and knew when to defer to his congressional peers. His Vice President, a stocky, headstrong man of forty-two, was, in personality and political ideology, his antithesis. The contrast in the style of these two presidents, as their countrymen would soon find out, was the difference between a benediction and a Rough Rider call to action.

Roosevelt had been a New York legislator, North Dakota rancher, reform-minded Civil Service Commissioner, brigade commander in Cuba, and Governor of New York.

In outlook as much a Dakotan as a New Yorker, TR was an outdoorsman who loved the land and gloried in "the strenuous life." Wildlife and big game were his first love, and in his early years he had published three books on his experiences as a hunter, naturalist, and rancher. His conservation interests did not range as far as Pinchot's, and the subtleties that fascinated Muir eluded him, but he was acquainted with the grass and water and soil of the Great Plains, and had sharpened his larger insights by writing a frontier history, *The Winning of the West*.

Nor was TR a stranger to the fight for conservation. In 1888 he had formed the Boone and Crockett Club—an elite and influential group of "American hunting riflemen" who fought the ruthless wildlife slaughter by commercial hunters, and persuaded Congress to make the Yellowstone country a refuge for buffalo and big game. As Governor of New York, he had hammered out new forest policies with the help of Pinchot, and had taken a keen personal interest in the management plans of his game and forest wardens.

Senator Mark Hanna and the McKinley men had been uneasy about TR from the moment he was thrust upon them at the 1900 Republican convention. That "damned cowboy," as he was later called, had a maverick look about him, and the old-line politicians could scarcely conceal their distrust and dismay.

But in the curious campaign of 1900, "The Colonel" had allayed some of their misgivings by the zest of his jingoistic stumping as he matched the indefatigable William Jennings Bryan speech for speech across the land. TR wholly ignored, in several hundred appearances, the question of the power of the "industrial overlords," and the resource-use issues which, fourteen months later as President, he would say were "the most vital internal problems of the United States."

When McKinley was assassinated, six months after the inauguration, his successor was called down from a wilderness camp in the Adirondacks to become the youngest President in our history. Roosevelt had already drawn heavy draughts from the well of political chance, but accident had intervened again, and the overlords and their *laissez-faire* ideas of resource development were due for a direct challenge.

He welcomed the turbulent task of presidential leadership. Temperamentally another Andrew Jackson, he actively sought power, enjoyed controversy, and was privately disrespectful of his elders on Capitol Hill. Moreover, his convictions about the land ran counter to the prevailing view, and he was filled with a zeal and impetuosity that made him an obvious instrument of change.

The outstanding resource achievement of Theodore Roosevelt's first term was the Reclamation Act of 1902, which ended a long argument over whether the states, private groups, or the federal government should develop the rivers of the seventeen Western states. This bill was sponsored by Democratic Senator Newlands of Nevada, but TR helped round up the votes and gave it the extra push that put it through.

This Act created an agency to work on the water problems of the West and earmarked receipts from land sales and mineral royalties for a revolving fund to finance dams and canals. By limiting the sale of water from federal irrigation projects to farms of 160 acres or less, the Reclamation Act struck the first broad blow for land reform since the Homestead Act of 1862. Here were most of the essentials of the old Powell policy for the arid lands. By the time Roosevelt left office seven years later, thirty projects were under construction to reclaim 3,000,000 acres of rich bottomland, and a trained team of hydrologists and dam builders had been assembled that would, in time, win world-wide attention for its water work and land reclamation.

This was the kind of positive federal program John Quincy Adams had called for seventy years earlier. It remained the finest example of legislative-executive teamwork during TR's two terms. Although the Congress created five new national parks, and voted the necessary funds for an expanded forestry force, a majority of the legislators remained hostile or indifferent to the executive-action part of Roosevelt's crusade for conservation.

During TR's first term, the raiders continued their relentless pressure on our national estate. Roosevelt saw that radical countermeasures were needed, and the temper of his mind—self-righteous and cocksure—was such that he determined to throw all of the latent powers of the presidency into the fight the moment the time was ripe. In the interim, the hawk-eyed Pinchot surveyed the terrain, and once TR received a rousing mandate of his own in the election of 1904, a light-brigade charge for land conservation began.

In the four furious years that followed, Theodore Roosevelt rewrote the rulebook on presidential power, effectively reversed the nation's thinking on resources, and saved a land birthright for the American people.

Roosevelt was a success because of his concept of the presidential office and its powers. He regarded himself as the trustee of the lands owned by the people. It was his imperative duty, he was convinced, to stop "the activity of the land thieves" and reserve the best of our national lands for permanent public use. TR flatly rejected the "narrowly legalistic view" of his predecessors that he could function only where a statute clearly bade him to act. He was the servant of the people, not of the Congress, and the charter he looked to for his power was the Constitution itself. Furthermore, he proposed to function fully and affirmatively as the nation's landlord and chief husbandman wherever and whenever his sense of stewardship told him action was needed. Under his touch (to paraphrase Mr. Dooley's memorable quip about corporation lawyers) the solid statutory walls of the past were turned into the triumphal arches of his administration. Most congressional leaders thought the performance arrogant, but it was a positive hour for the public good.

As Roosevelt spoke eloquently of the people's stake in their resources, and of the duties owed one's children and grandchildren, he put the landgrabbers on the defensive. Time after time he denounced "empire builders," who were concerned only with amassing personal profits and in the process were destroying the common heritage.

TR succeeded because he dared to use his pen. The

series of orders and proclamations that flowed from the White House dedicated millions of acres of public lands for forests, parks, and wildlife refuges.

His methods were often unorthodox. In 1908 a canny speculator (later a United States Senator) saw the tourism potential of Arizona's Grand Canyon, and sought to control access to it by preempting the awe-inspiring overlooks with a series of mining claims. The move would have been entirely legal, as the Canyon was part of the public domain. The problem was taken with alarm to the President. His first impulse was to make the Canyon a national park, but there was not time to wait for an Act of Congress. Two years earlier, however, Congress had passed the Antiquities Act, giving Presidents the power to create "National Monuments" for the preservation of "historic landmarks . . . and other objects of historic or scientific interest." Although the Grand Canyon seemed scarcely to fit the description, the Attorney General told Roosevelt he could take a chance and make it a national monument. TR signed a proclamation on the spot, and withdrew the canyon and its rims from mining entry. Twelve years later (after Roosevelt's death) the United States Supreme Court, in a broad-minded moment, did some conserving of its own and upheld the action, putting the indelible ink of authority into a pen that congressional committees had hardly intended to place in presidential hands.

Then there was the story (perhaps apocryphal, but highly illustrative) of Washington's worst eyesore—the old Pennsylvania Railroad station on the Mall. Roosevelt wanted it moved. Advised that he lacked legal authority to move it, and that no statute could be found which would permit him to act, he inquired artfully if the Attorney General had found any statute that prohibited him from razing the station. The reply was negative, and the President ordered the offending structure torn down forthwith.

TR proved that the pen was mightier than the legislative act, and before he finished he had, in effect, replaced the century-old policy of land disposal with a new presidential policy of withdrawal-for-conservation. National forest lands were increased from 42,000,000 acres to 172,000,-000 acres, and 138 new forests were created in 21 states; he carved out 4 extensive wildlife refuges, and withdrew 51 smaller reserves for birds and waterfowl to protect "the beautiful and wonderful wild creatures whose existence was threatened by greed and wantonness"; in addition, he proclaimed 18 national monuments, including 4—Grand Canyon, Olympic, Lassen, and Petrified Forest—so majestic that Congress later made them national parks.

Another Roosevelt project, masterminded by Pinchot, was the levying of fees on stockmen whose herds grazed on forest lands. No law directed the charging of such fees, and the measure stirred up a storm in the West, but TR and GP operated on the stewardship-landlord principle that those who used public property should pay for what they got, and this big step marked the beginning of modern management of the public lands.

In the final phase of the Roosevelt administration, largely under the influence of Pinchot, the forest-conservation idea was broadened to include minerals, hydro-power, and all other natural resources. When the President vetoed two bills that would have turned over choice hydro-power sites to private companies, he realized that a use plan was needed for all resources. Withdrawal, by itself, was not a sufficient policy. TR wanted Congress to establish leasing systems that would prevent monopoly and protect the public interest; he forced action by a sweeping series of orders that reserved the best hydro-power sites on sixteen Western rivers, as well as nearly 75,000,000 acres of public coal and phosphate lands. There was a gnashing of teeth on Capitol Hill at this new "lock-up" of resources, but his action ultimately had the desired effect: after a twelve-year wrangle, Congress, in the last year of the Wilson administration, enacted two basic conservation measures—the Mineral Leasing Act of 1920, and a bill establishing a Federal Water Power Commission. Under public-trust principles, fees and royalties were fixed, and hydro-power sites were made eligible for licenses under terms which protected the public interest.

The culmination of Roosevelt's effort to re-educate his countrymen was the White House Conference on Conservation he convened in the spring of 1908. This presidential Chautauqua did more to crystallize opinion than any event of TR's tenure. It was salesmanship of the highest order, and it consecrated Pinchot's new watchword—conservation. In a keynote address to an assemblage that included most of the leading political figures of the day, the President outlined his land philosophy and rejected the creed of the past which had allowed "the individual to injure the future of the Republic for his own present profit." He argued that it was time for resource planning by trained men, for employing the full power of federal and state governments to prevent waste, and for beginning the co-ordinated development of all resources. A few of the Western governors grumbled about states' rights, but for nearly a week, under presidential tutelage, the nation had an unsurpassed schooling in public affairs.

As Theodore Roosevelt's crusade widened, it became a quiet revolt of the American middle class against the waste of the raiders and the excessive power of those busily building economic and political monopolies. In its final phase, the fight for resource conservation became a larger fight for what TR called "real democracy." Big business was confronted with a big government which insisted that the common good take precedence over individual ambitions, that easy profits take second place to the betterment of society, and that our resources be managed under rules of prudent stewardship. The Roosevelt-Pinchot campaign refreshened the idea of democracy itself.

In the years following the turn of the century, the nation needed new direction, and TR was at his best as a teacher and preacher-at-large. The country had to be aroused, and the idealistic young President was an arouser whose revivalist zeal carried the crowd with him.

Where conservation was concerned, TR was not bashful about brandishing his big stick. Nor did he speak softly. A group of foresters never forgot how he threw away a written speech and roared, "I hate a man who skins the land." His denunciations of soft living and "the get-rich-quick theory of life" had the force of an Old Testament sermon. The animating thrust of his personality still burned brightly in the mind of his friend, William Allen White, when, in his old age, he recalled the tingling impact of their first encounter:

> . . . It was out of the spirit of the man, the undefinable equation of his identity, body, mind, emotion, the soul of him, that grappled with me and, quite apart from reason brought me into his train. It was youth and the new order calling youth away from the old order. It was the inexorable coming of change into life, the passing of the old into the new.

TR's influence did not end when he left the White House in 1909. His concept of conservation finally became fixed as a part of the American creed during the course of the rousing political campaign of 1912. Once TR reentered politics, it was certain that his opponents would have to compete with him as champions of the conservation cause. This made the campaign one of the most provocative contests in our history, and many of the men who were destined to lead the nation in the troubled days of the 30's —men like Ickes, Norris, Rayburn, Stimson, and FDR himself—lit their political torches at the great bonfire of 1912. TR failed to regain the White House, but his Bull Moose effort provoked a debate that educated the American people as they had never been educated before. Thereafter, conservation and democracy were a single, inseparable creed.

Theodore Roosevelt dealt a decisive blow to the Myth of Superabundance, slowed the raiders to a walk, and gave us new attitudes toward the land and a new appreciation of the nature of democratic government. His approach, however, had shortcomings, the most serious of which was related to the inherent limitation of his presidential stewardship policy. Executive action worked well as far as it went, but was essentially a policy to save what was left of the Western lands. Large-scale legislative action would be necessary to renew the vast areas of our continent damaged by the Big Raids. Roosevelt's tendency to carry the ball himself encouraged the public to adopt a let-Teddy-do-it attitude toward conservation. Congressional lethargy was underscored by the fact that in the decade following 1909 only two far-reaching conservation measures—the Weeks Act authorizing a system of Eastern national forests, and the National Park Service bill of 1916—were enacted.

No one realized yet the full scope of the conservation challenge: the task of land rehabilitation would be a long, costly process, and it would take a full-time partnership between the executive and the Congress to do the job right. Further, it would take bitter years of depression and defeat to drive the point home.

During the interval between the two Roosevelts, conservation moved forward in some areas, but on the whole our land continued to lose its vitality. The Reclamation Service was hard at work bringing water to farms in the Western valleys, but there was no organized program to renew the soils of the farmed-over lands of the East. The "wheat rush" of World War I had encouraged the improvident plowing of vast grassland tracts on the Great Plains and the Forest Service was spreading the sustained-yield gospel, but little was being done to reforest and repair the vast cutover woodlands; and although the Park Service was busy enlarging our scenic estate, wildlife habitat continued to shrink. One of the few constructive steps in this period was the Norbeck-Andresen Act of 1929 which became the basis of our national system of waterfowl refuges.

Both waste prevention and projects of positive development lagged far behind national needs. Although the Reclamation Service had built a number of irrigation projects, we rejected the advice of Powell and Pinchot and TR to make comprehensive plans for river-basin development, and the Hoover Dam project, the only big river-control structure that had been approved by the Congress, ignored principles of sound regional planning. Ninety-five per cent of the hydro-power of our rivers was unharnessed, and nearly all of our farm and ranch families were denied the benefits of electric power. Soil erosion, the primal form of waste, was a problem in the West where the public-domain grasslands were overgrazed and unregulated; and in the East the soil-saving and fertility-restoring practices of our farmers were either unsystematic or nonexistent.

The economic bankruptcy that gnawed at our country's vitals after 1929 was closely related to a bankruptcy of land stewardship. The buzzards of the raiders had, at last, come home to roost, and for each bank failure there were land failures by the hundreds. In a sense, the Great Depression was a bill collector sent by nature, and the dark tidings were borne on every silt-laden stream and every dust cloud that darkened the horizon. Land failure meant the failure of people, and in the early 30's mortgage foreclosures and a daily deposit of dust on some window sills underlined this lesson for those who lived in the mid-continent.

In the period which led up to this debacle, nothing depicted our indecision better than the fight over Muscle Shoals, a choice hydroelectric site near rich phosphate deposits on the lower reaches of the Tennessee River. The Muscle Shoals debate began in 1903 when TR vetoed a special bill that would have licensed the site to a private utility company. It flared up over the years and came to a head twice, in 1928 and 1931, when federal-development bills shepherded through the Congress by the dogged Nebraskan, Senator George Norris, were vetoed by Calvin Coolidge and Herbert Hoover. The latter drew a sharp issue for the 1932 campaign when he said the Norris bill would "break down the initiative and enterprise of the American people . . . [and was] the negation of the ideal upon which our civilization has been based. . . ." Franklin Delano Roosevelt gave public development of Muscle Shoals a ringing endorsement in his campaign, and raised the old banner of his cousin, Theodore, with his assertion that we lacked a land policy worthy of the name.

It would, in fact, take a much larger national effort than either TR or Pinchot had envisioned to arrest the land devitalization dramatized in the early 30's by the damaging floods of our major rivers, the spreading blight of rural poverty, and, finally, the sickening spectacle of dust arising from the Great Plains.

We learned, but the learning process was slow, and in the main we waited for the shock of man-caused disasters to awaken us to the full failure of land stewardship. When, in 1933, the second wave of the conservation movement belatedly arrived, under the second Roosevelt, it elicited enthusiastic congressional participation, a dozen or more new agencies and services, massive appropriations, and a series of new resource programs.

Roosevelt was superbly prepared to lead the new conservation campaign. Standing on a platform built by his predecessors, he could see clearly the mistakes of the past, and he realized that one of the best ways to galvanize a demoralized people was to institute programs that would renew and rehabilitate the land.

Conservation was more than a political creed to FDR. He cared about the continent, and his mind was bubbling with ideas about land and water and wildlife. Nurtured on TR's concepts, he had started his public service in the New York Senate in 1910, and once described himself in *Who's Who* as "a tree grower." For years he had personally supervised the rebuilding of the topsoil on his erosion-scarred farm at Hyde Park and had planted as many as 50,000 trees on his estate in one year. The forests, he once said, are "the lungs of our land, purifying our air and giving fresh strength to our people."

As New York's depression governor he had sent 10,000 unemployed men into the state forests to plant trees, and in his 1932 acceptance speech at the Democratic Convention he foreshadowed his approach to conservation with this piece of Bull Moose oratory:

> Let us use common sense and business sense; and just as one example, we know that a very hopeful and immediate means of relief, both for the unemployed and for agriculture, will come from a wide plan of the converting of many millions of acres of marginal and unused land into timberland through reforestation. There are tens of millions of acres east of the Mississippi River alone in abandoned farms, in cutover land,

now growing up in worthless brush. Why, every European nation has a definite land policy and has had one for generations. We have none. Having none, we face a future of soil
erosion and timber famine. It is clear that economic foresight
and immediate employment march hand in hand in the call
for the reforestation of these vast areas.

He proposed that we put one million men to work in the
out-of-doors conserving wood and water and soil. President Hoover's Agriculture Secretary hooted at the suggestion, but FDR was confident his idea was sound, and his
election assured a tryout for a conservation corps.

When the New Deal began, conservation became an integral part of the war against depression. Had FDR selected a slogan, it might well have been "Look to the land."
During his first term, more lasting land-renewal work was
accomplished than during the previous thirty, and the job-
less men who did the task found their self-confidence restored as well.

This was a new kind of American empire building; its
goal was the common good; its familiar alphabetic emblems were CCC, TVA, AAA, and SCS; and its most visible
symbols were the huge dams that began to rise from the
bedrock of such rivers as the Columbia, the Tennessee, and
the Sacramento. To FDR, our thinking had to be continent
wide, and our resource planning all-inclusive if we were
to avoid the "haphazard and piecemeal" development deplored by Pinchot and TR. The price of a permanent prosperity was the wise use of all of the land and its products.
Unity was his keynote, for "The east has a stake in the
west, and the west has a stake in the east . . . [and]
the nation must and shall be considered as a whole and
not as an aggregation of disjointed groups. "

The first conservation fruits of the New Deal were two
of the big bills of the famous One Hundred Days, which es-

tablished a Tennessee Valley Authority and a Civilian Conservation Corps. Under the first, the old Muscle Shoals proposal was transformed into a broad plan for the redevelopment of a region that included parts of seven states, and a semi-independent regional "authority" was created and given wide powers to promote "the economic and social well-being of the people" of the entire valley.

The Tennessee Valley had been rich in timber and petroleum resources, but the raiders had come and gone, leaving a demoralized people to pick up the pieces. No region of the United States had lower incomes or more families on relief. When FDR spoke of "one third of a nation ill-housed, ill-clad, ill-nourished" he might well have had these people uppermost in his mind.

By 1933 this region of thin soils had become one of the most depressed and most depressing areas in all the United States. "Three fortunes had been taken off that country—forests, oil and gas," said one of the original TVA Directors, Dr. Arthur E. Morgan, who wryly observed that "The wreckage of the rugged individualism has been handed to us with a request that we try to do something about it." A final piece of mockery made matters worse, for the Tennessee River had a score of superior hydro-power sites and yet its silt-swollen waters rushed unharnessed and unchecked through counties and states where very few farmers had electricity. The TVA dam building soon began. In addition, the FDR plan included navigation, reforestation, soil conservation, outdoor recreation, the retirement of marginal farmland and the long-discussed manufacture of fertilizer.

In a decade, the river was put to work, over 700 miles of streams were made navigable, 200,000,000 trees were planted in the uplands, the dams provided abundant low-cost power, and scientific agriculture finally got a foot in the door. Comprehensive "source to mouth" river-basin planning was begun on the largest scale in our history, and the economy of the region was revitalized as cheap power attracted new industries and created jobs for the unemployed.

As the experiment got under way, shrill voices shouted "socialism," but the heirs of Theodore Roosevelt had invented an economic and social "ism" that was as indigenous and homespun as Senator George W. Norris, who had sponsored the TVA bill and had fought for the land renewal of the Tennessee Valley.

TVA was a triumph of planning such as Theodore Roosevelt and Pinchot had envisioned. Some admirers have termed it, "the greatest peace-time achievement of Twentieth Century America." It raised a beacon of hope for many Americans, and is still the resource conservation project most often visited by government leaders of other countries.

TVA helped open the door to similar, but less systematic, development work in other river basins: the Reclamation Bureau commenced work on two giant main-stem dams on the Columbia (Bonneville and Grand Coulee); dam and canal building raced forward in the Central Valley of California; and on the Missouri, the Arkansas and other large rivers, ground was broken for major projects.

A few ill-planned efforts such as the Passamaquoddy tidal project in Maine were undertaken, but FDR generally was proud of his "national plan," and on reviewing the results of his conservation programs he exuberantly told reporters, "You might almost call it rebuilding the face of the country."

With large amounts of federal hydroelectric power soon to be ready for marketing, it was logical that the New Dealers would turn their attention to another area of gross neglect—rural electrification. Of the 30,000,000 rural Americans in 1934, nine-tenths were without electrical service. Through default of the private power industry, the age of light had not reached the farms and ranches, and it was plain that federal help was needed. Congress, in 1935, underwrote the formation of rural electric co-operatives and provided loans for transmission lines, and within fifteen years the new program had brought electrical energy to nine-tenths of America's farms. This was a striking achievement of comprehensive planning, and the Rural Electrification Administration was a bringer of light to the men and women in the countryside and a lightener of the burdens of their everyday lives.

If Senator Norris was the godfather of the TVA, Franklin Roosevelt himself was both father and godfather of the CCC. Forestry was his avocation and he once told Harry Hopkins that every boy should have the opportunity to work for six months in the woods. This was the happiest of the New Deal programs, for it simultaneously rehabilitated land and men. In essence, it was a broad effort to repair the worst of the damage done by our raid and waste policies. Corpsmen built small dams, tackled erosion problems, planted over 2,000,000,000 trees, aided wildlife restoration, and built needed facilities in the National Parks. Before World War II closed the camps, more than 3,000,-000 young men served in the Corps, and the top enrollment reached over 500,000 in 1935. The CCC was essentially an investment in the future, and it received an accolade a generation later when President Kennedy proposed that it be revived to resume the work of land rehabilitation.

In another bold stroke for forest conservation, FDR gave orders for a vast expansion of the Eastern forests by the purchase of cutover land. From 1911 until the New Deal began, twenty-four national forests had been created under the Weeks Act, but only 5,000,000 acres of land had been actually acquired. In less than thirty months, Roosevelt allocated over $37,000,000 of public works monies to purchase 11,000,000 additional acres, and the restoring work of the CCC was immediately commenced.

Inherent in the New Deal approach to resources was an assumption which FDR did not spell out, but which was the heart and soul of his conservation effort. During the era of the Big Raids, much of the nation's resource capital had been borrowed and "used up" to advance the personal fortunes of the few. The New Dealers, in effect, reversed this prodigal process, and borrowed from the future (by deficit financing) to invest in land-rebuilding programs that would assure adequate resources for tomorrow. With this approach, the needs of the community and of the next generation were given first priority.

Nowhere were these needs more evident than on the western border of the Great Plains where nature struck back and made us add up the cost of our bankrupt land policies of the past. "The topsoil in dry seasons is blown away like driven snow" FDR once declared, and Agriculture Secretary Henry Wallace wondered aloud whether we were not greater soil wasters than the Chinese. A systematic, nationwide effort was needed. The farmers and landowners had to be organized for action.

In 1935 a Soil Conservation Service was created, and to head it Roosevelt picked a modern land evangelist, Hugh Hammond Bennett, who quickly made his countrymen soil-conscious. For the first time soil and moisture conservation was part of the country's urgent business and we began to look on discolored streams as a sign of failure. Bennett estimated that the annual cost of erosion was at least $400,000,000 in terms of diminished productivity alone.

In 1934, Congress finally took action on the most neglected part of our national estate—the grasslands of the West. The Taylor Grazing Act effectively closed the public domain to homesteading, established grazing districts, and set up a program of regulated use to control overgrazing and prevent erosion.

The President was also keenly interested in wildlife. He issued memorandums on behalf of wild things like the Grand Teton elk herd and the trumpeter swan, and had a running correspondence with "Ding" Darling (the noted cartoonist who briefly joined the Administration) over the plight of our waterfowl. In 1937, Congress passed the Pittman-Robertson Act, which levied a tax on firearms and ammunition to provide monies for state wildlife projects.

Nothing better illustrates FDR's intimate interest and flair for dramatic active intervention than an incident that occurred a week before the attack at Pearl Harbor. He had received protests from conservationists that a proposed army artillery range at Henry Lake, Utah, might result in the extermination of the trumpeter swan.

Roosevelt responded with this memorandum for the Secretary of War:

> . . . Please tell Major General Adams or whoever is in charge of this business that Henry Lake, Utah, must immediately be struck from the Army planning list for any purposes. The verdict is for the Trumpeter Swan and against the Army. The Army must find a different nesting place!

FDR believed it was necessary to put "the physical development of the country on a planned basis," and he created a National Resources Planning Board to make a comprehensive analysis of all resources and to frame plans for their use and development. This was a brief hour of triumph for scientific planning. Before the anti-planners in the Congress cut off its funds, the board filed a series of exhaustive studies that are still bench marks of conservation planning.

By 1935 the new campaign for conservation was in high gear. As the outdoor brigade did its work, old scars were gradually healed, and the slow process of land recuperation began. Millions of Americans now looked on our resources with a new sense of personal responsibility and gave fresh support to the effort for national stewardship.

The words and deeds of the Roosevelt eras are an eloquent legacy. We owe a large debt to those who crystallized and enlarged the conservation concepts that originated with Thoreau, Marsh, and those who followed. The challenge of our own time is to adapt this creed to the diverse and bewildering problems of a more complex age.

CHAPTER XI

Individual Action:

ORGANIZERS
AND PHILANTHROPISTS

"Who would not rise to meet the expectation of the land?"

—HENRY DAVID THOREAU

As the master politician navigates the ship of state, he both creates and responds to public opinion. Adept at tacking with the wind, he also succeeds, at times, in generating breezes of his own. Both Theodore and Franklin Roosevelt stirred up winds of public opinion, but they could not have prevailed without the toil and sweat of individuals and organizations who did yeoman service for the conservation cause.

Any land history of this country would be incomplete without an account of the contribution of the land philanthropists and our conservation organizations. Just as the search for a sound land policy is a quest for the right balance between public and private ownership, so our political system is an ongoing endeavor to find the right combination of government action and private effort. To a large degree the conservation history of this century is the story of the creative interplay of individuals, associations, and the agencies of government.

The voluntary movement began a few years after Thoreau and Marsh lodged their complaints with the public conscience. In 1875 a small group of perceptive men, dismayed by the reckless invasion of the forests, established the American Forestry Association to protest waste and to devise ways and means of saving the remaining forest land. The AFA supported Carl Schurz's initial efforts for reform, and was in the forefront of the long campaign that led to the Forest Reservation Act of 1891. Conservation acquired another powerful voice in 1885 when a group of New York ornithologists, led by George Bird Grinnell, formed an association to protest the commercial hunting of birds and the senseless slaughter of wildlife. The group became the nucleus of the National Audubon Society, which soon became involved in a campaign against the use of bird feathers on women's hats. This campaign, led by William Dutcher and T. Gilbert Pearson, was a bit sentimental at first, but it gathered public support and altered American attitudes toward wildlife.

The crusade of the Auduboners bore legislative fruit in 1913 when a tariff act prohibited the importation of feathers. This was a blow to the milliners, but gave a new lease on life to such birds of bright plumage as the egret. Other triumphs of the conservers of birdlife were the adoption of migratory bird treaties with Canada and Mexico and the establishment of a system of refuges for waterfowl.

Grinnell, also a big-game hunter, zoologist, and friend of the Plains Indians, was in the 1880's the editor of *Forest and Stream Weekly*, which carried on a crusade for wildlife and for better rules of sportsmanship in hunting. One December night in 1887, at a dinner in Manhattan, a youthful Theodore Roosevelt proposed to Grinnell and a group of hunters that they form an organization to fight the indiscriminate slaughter of big game.

With a bow to their mentors of the old frontier, they named their elite group of "hunting riflemen" the Boone and Crockett Club and set a membership limit of one hundred. Together they possessed a prestige and breadth of experience that gave them entree to offices of influential men. They set the standards of sportsmanship for a generation and did much to save the big mammals of North America. No conservation organization in our history has had more political know-how, and they soon succeeded in getting a law passed prohibiting interstate shipment of the meat of wild animals.

As hunters with an active interest in biology, Boone and Crockett members knew the importance of animal habitats, and they turned their attention to the management of the public land. They secured authority for the wardens of Yellowstone Park to prosecute game poachers and persuaded Washington officials that small military units should be assigned to protect parklands. The Boone and Crockett wildlife creed soon became national policy when Theodore Roosevelt became President, and a system of big-game refuges was begun in 1905 when he established the Wichita Mountains refuge in Oklahoma. For Boone and Crockett members, the bison was the symbol of the West, and they made a *cause célèbre* of the buffalo massacre. TR later gave them—and posterity—a Montana refuge where the shaggy survivors could begin a comeback from the edge of extinction.

The Boone and Crockett Club has made an outstanding contribution to our legacy of wild things. Its achievements measure up to the best efforts of our private land philanthropists. Chief among them was John D. Rockefeller, Jr.,

son of the Standard Oil magnate. John junior spent most of
his life presiding over the creative redistribution of a
family fortune realized from the fossil-fuel resources of
the American earth. The elder Rockefeller had become the
Paul Bunyan of monopoly capitalism. At the very pinnacle
of his career, his adviser, Frederick Gates, warned, "Your
fortune is rolling up, rolling up like an avalanche! It will
crush you and your children, and your children's children."

John junior took this warning to heart and set out to
channel the family wealth into philanthropic activities. A
mild-mannered, abstemious man, the young Rockefeller
from childhood had had a feeling for nature and a love for
fine landscapes, and as the years went by, large pieces of
the Rockefeller fortune were invested in conservation
projects.

His initiation in saving parkland occurred at the site of
his family's vacation retreat on Mt. Desert Island, the
scenic gem of down-East Maine. With Rockefeller as un-
derwriter, a group of residents established the Hancock
County Trustees, which acquired through purchase and
gift some of the island's best landscapes, including the
summit of Mt. Cadillac, where the morning sun first
touches the American headlands. In 1916 a nucleus of
5,000 acres was presented to the United States and proclaimed by President Wilson as a national monument.
After subsequent additions, Congress designated the area
as Acadia National Park—the first national park east of the
Mississippi, and the first established solely by land donors.

Rockefeller designed for the park a system of carriage
roads that gently curved around headlands and pond-filled
hollows. If hand labor was required to carry out nuances of
design, he ordered hand labor, and when automobile roads
became a necessity, he brought Frederick Law Olmsted, Jr.,

to Acadia to guide the Park Service in developing a roadway system to fit the island's unusual geography.

The Acadia experience led Rockefeller into other conservation activities. At the turn of the century, when the bold Palisades of the Hudson River were being relentlessly quarried, his money made possible the purchase of the operating sites and helped to inspire a series of donations from others. As a result, thousands of acres in the Hudson highlands were secured for a regional park. Over the ensuing decades, Rockefeller continued to purchase palisade overlooks until over $20,000,000 had been invested in a broad, unspoiled park for the harried citizens of nearby New York City.

John junior also developed the technique of using his funds as conservation seed corn. In 1926 Congress had provided that land-acquisition funds for the Great Smoky Mountains and Shenandoah national parks were to be raised by public subscription; when local coffers failed to fill up, Rockefeller matched money with the states of North Carolina, Tennessee, and Virginia to save the Blue Ridge high country that has become the most-visited national park in the United States.

On a 1924 trip in Yellowstone Park, superintendent Horace Albright communicated to Rockefeller his own love for Jackson Hole—a wintering valley of the mountain men—below the bold rampart of the Grand Teton Mountains of Wyoming. Told of previous failures to extend Yellowstone Park to include the valley, Rockefeller became convinced that the purchase of Jackson Hole for a park would be an "ideal project," and in 1927 he formed the Snake River Land Company to acquire quietly the ranch property at the foot of the jagged peaks. Some ranchers were anxious to sell; others were irate at the idea, and their Wyoming Congressmen set out to thwart this "Eastern plot" by opposing any park legislation for the area. By 1943 a stalemate existed. At this point Interior Secretary Harold Ickes and FDR took a leaf from Theodore Roosevelt's book and bypassed Congress by proclaiming the area covering the donated Rockefeller lands and federal forests to the West a national monument. Congress, observing its rule of local sovereignty in land matters, passed a Wyoming-sponsored bill which would have abolished the monument, but FDR stood his ground and vetoed the bill, and a decade later a further compromise led to the establishment of the Grand Teton National Park.

For several decades the huge Rockefeller fortune contributed to the national estate. During the depression and World War II, when public funds were largely unavailable, John D. junior tipped the scales and helped to add several new parks to the heritage of his countrymen.

Rockefeller's philosophy of land philanthropy has been shared by such men as former Governor Percival P. Baxter of Maine. When he failed as a member of the legislature and as governor to secure state funds to save the spectacular landscape surrounding Mt. Katahdin, he took on the job as a personal project. By Rockefeller standards, Baxter's wealth was modest, but with a bit of New England thrift, he purchased piecemeal, between 1931 and 1962, a scenic area of 202,000 acres. In 1962, on the hundredth anniversary of Thoreau's death, he carried out the dream of a lifetime and presented the final tract to the state. Baxter State Park is today the most majestic state park in the nation. The donor attached one condition to his gift, the requirement that this region of moose and beaver country, of lakes and granite mountains, remain for all time inviolate from the mechanical intrusions of man. At a little nook called Thoreau Spring there is now a bronze plaque which recites Percy Baxter's creed:

Man is born to die. His works are short-lived. Buildings crumble. Monuments decay, wealth vanishes, but Katahdin in all its glory forever shall remain the mountain of the people of Maine.

Another philanthropic conservationist, Boone-and-Crockett-member Bayard Dominick, walked unannounced into the Washington office of Director J. N. "Ding" Darling of the Biological Survey one day in 1934 and offered his countrymen a gift of his family estate at Bull's Island, South Carolina. Darling accepted with alacrity, and this gift became the nucleus of the Cape Romain National Wildlife Refuge, which today embraces nearly 35,000 scenic acres of waterfowl and wild turkey habitat.

Many states have benefited from parkland donations, and the city of Detroit has a gift necklace of 14 parks which encircle it. Of the first 64 Michigan state parks, 59 were gifts. Hundreds of natural reserves across the country —from Taconic State Parkway and Bear Mountain in New York, to Calaveras Grove and Humboldt Redwoods in California—have resulted from landscape donations by generous men and women. Various organizations have, over the years, equalled or excelled the land philanthropists in the legacy they have left us.

When Steve Mather was called to Washington in 1915 to take charge of the national parks, he turned to one of his old associates, Robert Sterling Yard, for help in educating the American people about the national-park idea. Yard joined Mather in the Interior Department, but after passage of the 1916 National Park Act, he left the government and formed the National Parks Association to be a permanent voice for the parklands.

He astutely selected "men of high horizons" for his membership roster, and the organization made an impact on the Washington scene. Its first serious test came on the proposed Federal Water Power Act. Mather and Yard both knew from the Hetch Hetchy fight that reservoir proposals were a constant threat to the parks, and both sought to exempt the parks from the provisions of the Act. Each, using different techniques of political persuasion, won their exemption, and the NPA has been the watchdog of Park standards ever since.

The National Parks were administered under wilderness protection principles, and another private association to protect our primeval landscapes had its origins in the work of Aldo Leopold, a farsighted United States forester. In 1924, Leopold secured protection for a large wilderness section in the Gila National Forest of New Mexico, and in the following year wrote an eloquent plea in *American Forests* magazine for preservation of the remotest large wilderness tracts throughout the country. Partly as a result, beginning in 1929, a system of national forest primitive areas was established.

It soon became evident that protection and extension of a system of wilderness zones would need strong private support. In the summer of 1934 a small group of conservationists met in Tennessee and formed the Wilderness Society. The crisis that brought them together was concern that congressional enthusiasm for sky-line drives would result in invasion of the best unspoiled areas in the East. The dominant figure at the conference was another forester, Robert Marshall. As a result of the leadership of Marshall and his colleagues, the Wilderness Society has helped to keep alive the public interest in the wilderness values emphasized by John Muir.

The Wildlife Management Institute whose history, under various names, dates back to 1911, has conducted research that has been largely responsible for the training of the managers and biologists of today's wildlife. In the 1930s' the National Wildlife Federation came into being to correlate the activities of thousands of local sportsmen's clubs which were the Boone and Crockett Clubs of grass-roots America. The federation has helped to educate a whole generation in the use of firearms and the tenets of conservation. Another noteworthy wildlife society is the Izaak Walton League of America, which, since 1922, has befriended a variety of wildlife and natural resource projects.

In its later phase, the Audubon Society has become one of the few conservation organizations to go into the land-management business. It maintains over a million acres in its various sanctuaries for rare-bird species, such as the whooping crane and the flamingo. Special crusades by this society have been launched against oil pollution on the high seas; the pollution of streams, rivers, and estuaries; and the improper use of pesticides. The flight of a bird is the moment of truth to an Auduboner, and the broadening base of his interests is also a hopeful omen.

Woods, as well as wildlife, have been preserved by other crusading organizations. Dozens of priceless redwood groves in California have been saved by the timely fund-marshaling of the Save-the-Redwoods League. The National Geographic Society, too, deserves special recognition for its superior educational effort on behalf of nature and her wonders, and its philanthropic work for the saving of special places. The world's tallest living thing, the Founders Tree, now memorializes the world-wide solicitude aroused by such men as Henry Fairfield Osborn, John Merriam, and Madison Grant.

Group philanthropy takes another form in the Nature Conservancy, a nationwide semi-scientific society with guidance from conservation-minded businessmen. The Conservancy is dedicated to locating and purchasing small, quiet nooks typical of wild America.

The majority of Conservancy projects are the result of testamentary bequests. It does not choose to be a land-managing agency, and as soon as legal safeguards are assured, it turns its holdings over to a government agency or a local managing organization. The Conservancy's accomplishments include some of the nation's finest cameo landscapes: the white-water and hemlock cathedrals of Mianus River Gorge, in Westchester County New York; the stately cypress of Corkscrew Swamp, Florida; the unique vegetation of Holly Ridge, Missouri; a prairie-chicken refuge in Illinois—and places with such fascinating names as Black Chasm Cave, Sunken Forest, Volo Bog, Wolf Swamp, Buzzardroost Rock, Hoot Woods, and Vernal Pool.

The Nature Conservancy method is an innovation that holds great promise for saving wilderness tracts. Men and women in many communities are helping to restore the balance in nature's bank by making well-considered conservation desposits. Another noteworthy example of this approach is the work of the Philadelphia Conservationists, who purchased the Tinicum Marsh on the outskirts of their city and have acquired other nearby seashore and woodland areas.

As our green space disappears, nature memorials become increasingly appropriate. In Washington, D.C., shortly after the death of Theodore Roosevelt, some of his friends bought an 88-acre wooded island in the Potomac and gave it to the country as a nature memorial to a man who had a lifetime love affair with the out-of-doors. Despite recent encroachments by a highway bridge and an inner garden of asphalt, many Washingtonians regard the woodlands of Theodore Roosevelt Island as the ideal memorial in our nation's capital.

The main flaw in the performance of many existing conservation associations is that most overconcentrate on a chosen holy grail, and too few organizations have entered the fight for the total environment. We sorely need negative-minded watchmen on the tower similar to the British organization of "Anti-Uglies," which devotes full time to the task of thwarting things unsightly and projects ill-planned. It should never be forgotten, however, that the real purpose of any holding action is to keep avenues open for positive activity. The slogan "conservation begins with education," expresses accurately the need for organized efforts—both in the schools and outside—to teach new respect for resources.

Healthy features of the voluntary associations are the variety of their roots and the wide range of their causes. The programs and interests of such organizations as the Sierra Club, the Garden Clubs of America, the farsighted and influential Outdoor Circle of women in Hawaii (which has kept our island state free from billboards) ; the Federation of Western Outdoor Clubs, the New York Zoological Society, the Conservation Foundation, Resources for the Future, National Recreation Association, Desert Protective Council, the American Planning and Civic Association, illustrate the range and vigor of contemporary effort. Some concentrate on purchasing land; some take research or education as primary goals; some combat billboards or plant trees; some are composed of hikers, cyclists, or canoeists—but together they have broadened our understanding of the American earth, and together they can form a rising chorus for the conservation cause.

Our resource problems in the 1960's are measured by the flyway of a bird, the length of a river, the half-life of an element, the path of a wind, the scope of the oceans, or the shape of our cities. The years ahead will require both public and private conservation statesmanship of a high order. The individual citizen and the voluntary association are the corpuscles of our political bloodstream—and over the long haul it is their day-to-day actions that will decide the legacy we leave for other generations.

Thoreau once said, "A town is saved, not more by the righteous men in it than by the woods and swamps that surround it. . . ." Few of us can hope to leave a poem or a work of art to posterity; but, working together or apart, we can yet save meadows, marshes, strips of seashore, and stream valleys as a green legacy for the centuries.

CHAPTER XII

Cities in Trouble:

FREDERICK LAW OLMSTED

Proud, cruel, everchanging and ephemeral city
To whom we came once when our hearts were high,
Our blood passionate and hot,
Our brain a particle of fire:
Infinite and mutable city, mercurial city,
Strange citadel of million-visaged time—
O endless river and eternal rock,
In which the forms of life
Came, passed, and changed intolerably before us!
And to which we came, as every youth has come,
With such enormous madness,
And with so mad a hope—
For what? . . .

—THOMAS WOLFE
From "The Ghosts of Time"

THE URBANIZATION of America has been a striking trend of the twentieth century. In Theodore Roosevelt's time we were still a predominately rural people; now we are predominately urban and we are become more so by the day.

Our cities have grown too fast to grow well, and today

they are a focal point of the quiet crisis in conservation. The positive appeal of the modern city—the stimulating pageant of diversity, the opportunities for intellectual growth, the new freedom for individuality—have been increasingly offset by the overwhelming social and economic and engineering problems that have been the by-product of poorly planned growth.

Under explosive pressures of expansion there has been an unprecedented assault on urban environments. In a great surge toward "progress," our congestion increasingly has befouled water and air and growth has created new problems on every hand. Schools, housing, and roads are inadequate and ill-planned; noise and confusion have mounted with the rising tempo of technology; and as our cities have sprawled outward, new forms of abundance and new forms of blight have oftentimes marched hand in hand. Once-inviting countryside has been obliterated in a frenzy of development that has too often ignored essential human needs in its concentration on short-term profits. To the extent that some of our cities are wastelands which ignore and neglect the human requirements that permit the best in man to prosper, we betray the conservation ethic which measures the progress of any generation in terms of the heritage it bequeaths its successors. The citizens of our cities must demand conservation solutions based on the principle that space, beauty, order and privacy must be integral to its planning for living. As long as those designers and planners who might help us create life-giving surroundings remain strangers at the gates we will not create cities which are desirable places to live. Today, "progress" too often outruns planning, and the bulldozer's work is done before the preservationist and the planner arrive on the scene.

Between 1950 and 1959, while our cities' populations increased by only 1.5 per cent, the population of our suburbs increased by 44 per cent. Even this flight to the suburbs—in part a protest against the erosion of the urban milieu—has had its element of irony, for the exodus has intensified our reliance on the automobile and the freeway as indispensable elements of modern life. More often than not, the suburbanite's quest for open space and serenity has been defeated by the processes of pell-mell growth.

Many mental-health experts have offered evidence of the corrosive effects on the human psyche of the unrelieved tension, overcrowding, and confusion that characterize city life. There is a real danger that the struggle with ugliness and disorder in the city will become so all-consuming that man's highest and most specifically human attributes will be frustrated.

The prime business of those who would conserve city values is to affirm that such human erosion is unnecessary and wasteful; that cities can be made livable; that with proper planning the elements of beauty and serenity can be preserved.

Urban America has had, in Frederick Law Olmsted, its own conservation prophet and master planner. Lewis Mumford once called Olmsted "one of the vital artists of the 19th Century"; he has had no peer in the United States as a community designer.

Olmsted did his pioneering work in a period when the need for public playgrounds was not recognized and the art of city planning was largely ignored. In 1859, while Thoreau was noting in his journal that each town should have a miniature wilderness park ("a primitive forest of five hundred or a thousand acres where a stick should never be cut for fuel, a common possession forever"), Olmsted was already developing a design for a proposed "central Park" in the heart of Manhattan Island, and thus began his career as a conserver of higher values in the city.

The key to Olmsted's genius was his insistence that all

land planning had to look at least two generations into the future. His versatile talents found many outlets. For a time during the Civil War he headed up the United States Sanitary Commission, the forerunner of the American Red Cross. As a temporary resident of California, before John Muir arrived, Olmsted played a leading role in the enactment of the Yosemite park bill signed by President Lincoln in 1864—and some of his park-management ideas anticipated the subsequent standards of the national park system. He did land-use planning for San Francisco, Buffalo, Detroit, Chicago, Montreal, and Boston. He was commissioned to landscape the grounds for the Capitol and the White House in Washington, and he designed the 1893 World's Columbian Exposition held in Chicago.

By orderly planning and provision for abundant natural areas, Olmsted believed cities could keep sufficient breathing and playing space to allow continual self-renewal. He proposed that part of the countryside be preserved within each city. His Greensward plan for Central Park was, as he later wrote, designed to "supply to the hundreds of thousands of tired workers, who have no opportunity to spend their summers in the country, a specimen of God's handiwork that shall be to them, inexpensively, what a month or two in the White Mountains or the Adirondacks is, at great cost, to those in easier circumstances. The time will come when New York will be built up, when all the grading and filling will be done. . . . There will be no suggestion left of its present varied surface, with the single exception of the few acres contained in the Park."

Central Park was a dreary stretch of rock and mud when Olmsted took charge. Working with nature, he tried to visualize and anticipate the growth patterns of a great metropolis. All effects in the park—trees, mounds, ponds, paths, meadows, groves—were carefully composed with an eye to creating life-promoting surroundings. Shrubbery screened out the works of man, and Central Park became an oasis where urban man could refresh his mind and soul.

Olmsted doggedly fought off the politicians and the well-meaning promoters who wanted to install on the grounds a stadium, a theater, a full-rigged ship, a street railway, a race track, a church, a permanent circus, a cathedral, and a tomb for Ulysses S. Grant. In each victory he affirmed the primacy of park purposes and strengthened the idea that some parkland had to remain inviolate.

As his vision broadened, Olmsted became more than a mere planner of parks. He saw that urban design should include the whole city and provide diverse and continuous enclaves of open space, green gardens, and public playgrounds. Had he been able to win support for his bold conceptions, the shape of many of our cities might be different today. His aim was to suit the city to the individual, and not vice versa, and perhaps his achievement of a healthy balance between the works of man and the works of nature in an urban setting is his most durable monument.

Despite the success of Central Park, the city fathers did not adopt Olmsted's most farsighted recommendations. Open space and elbow room cost money even then, and in a period of hectic growth the vision of a Frederick Law Olmsted was too advanced for the apostles of "progress."

The result was predictable. Like most of our large cities, New York fought a losing battle against congestion and blight, and Central Park today is a solitary symbol of what might have been. New York's failure to carry out Olmsted's plans meant that her second noted planner, the redoubtable Robert Moses (who began his public career in 1913) has had to concentrate largely on a costly and belated campaign to overcome earlier failures to plan. His beaches and parks have added immeasurably to the livability of New

York, but inevitably they have been insufficient to keep up with the needs of the millions. His energies, of necessity, have involved bold efforts to prevent paralysis of transport by the construction of new parkways, tunnels, and bridges. Where Olmsted placed his emphasis on grass and trees, on the re-creation of nature within the city, Moses has found himself involved to a great degree in working with asphalt for beach parking lots, for playgrounds, and for roads, including an abortive proposal in 1962 that a paved road be built through the remaining narrow, unspoiled stretches of Fire Island.

Jones Beach—an imaginative solution to the mass outdoor recreation problem of a megalopolis—is perhaps Moses' supreme answer to the ever-present problem of overcrowding. However, the long delay in implementing Olmsted's ideas meant that New York City had passed up the opportunity to develop a whole series of Central Parks and Jones Beach parks needed to provide an adequate outdoor environment.

Where, then, is the current battle line of conservation in our cities? What palliatives, what permanent remedies are available in the fight against congestion and decay?

It is significant that two of President John F. Kennedy's pieces of pioneering legislation have involved our cities: an Act providing financial aid to urban areas for the acquisition of open space and a proposal to assist cities in solving their mass transit problems.

The Olmsted ideas are still applicable, and even today most American cities have unrecognized opportunities, both within their corporate limits and on their fringes, to save large and small Central Parks for the future. Every well-conceived urban redevelopment project offers an opportunity to create green spaces in the central city and avenues of action are open to conservationist city leaders and citizen groups who set store by civic beauty and are willing to levy sufficient taxes for environment preservation. They should invite the Olmsteds of our time to participate in the re-designing of our cities.

No attitude is more fatal today than the belief of some local leaders that economic salvation lies solely in getting new property on the tax rolls. Central Park cost something over $5,000,000 in the 1850's. It is worth billions today, and much of its value lies in its ad valorem and esthetic enhancement of surrounding property. Money spent on a properly planned environment is an investment not only in future taxability, but in the physical and mental health of the residents—their efficiency, their general well-being, and their enjoyment of life.

If we are to create life-enhancing surroundings in both cities and suburbs, the first requirement is the power to plan and to implement programs which encompass the total problems of metropolitan regions. Air and water pollution, recreation, and provision for adequate mass transit are region-wide problems, but in most areas action is hampered by legal impediments which actually prevent regional plan-ning. As long as each city, county, township, and district can obstruct or curtail, planning for the future cannot be effective. The cities and metropolitan areas that are devis-ing new political institutions for regional planning are today's pioneers of urban conservation.

But local governments still hold the key to planning. Many zoning boards are as important as the courts. Zoning regulation should not merely prevent the worst from hap-pening . . . it should encourage positive action to provide esthetic opportunity for the present and future while pre-serving the history of the past. One example of creative zoning occurred recently in the Santa Clara Valley of Cali-fornia—a valley famous for its orchards, which unhappily were easy to bulldoze and presented an inviting target for developers working south from San Francisco. In the absence of adequate planning, checkerboard subdivisions made the remaining land difficult to farm and easy to exploit. Farmers were confronted by increasingly heavy taxes to finance the schools and services required by new residents.

Faced with the destruction of the valley's unique beauty, county planners and farmers evolved the idea of zoning the best orchard lands exclusively for agriculture. As they well realized, zoning is not a permanent answer to the problem, but it has saved many thousands of acres of farmland and provided breathing space for the subdivisions around them.

As the Santa Clara experience illustrates, an anachronistic tax structure which penalizes property owners who preserve open space or keep agricultural land in production is a serious impediment to sound land planning. The time has come for us to use the taxing powers of government as a creative force for conservation. Why not tax the owners of ugliness, the keepers of eyesores, and the polluters of air and water, instead of penalizing the proprietors of open space who are willing to keep the countryside beautiful? Open space and other socially beneficial land uses should be encouraged. The city of San Juan, Puerto Rico, has used novel tax laws to preserve the historic and picturesque "old city" by giving tax advantages to homeowners who remodel in conformance with the existing Spanish style architecture.

Both zoning and tax incentives may be used to stimulate another method of saving open space: cluster development. By clustering homes and designing smaller yards immediately surrounding them, a developer can provide the customary number of dwellings per acre and yet preserve as much as one-third of a given area in its natural condition.

This offers residents pleasant networks of wooded walks, stream-side parks and other recreation areas. This concept in a sense marks a return to the village greens and town commons of America's colonial period. Some communities in the United States and Canada have encouraged this kind of preservation of natural areas by requiring that a proportion of the total area in any new subdivision be dedicated to open space.

Another innovation in land planning is the conservation easement. In the seacoast county of Monterey in California aroused citizens secured the enactment of a state "open space" law in 1959, enabling cities and counties to purchase property, or easements on property, for the purpose of preserving pastoral landscapes. In purchasing an easement on open land, the public agency acquires a "right" from the owner, but otherwise leaves him full ownership and the property remains on the tax rolls. The "right" might be simply that the land remain in its natural state, as in the case of scenic easements. Usually an owner sells the right, but sometimes he may donate it voluntarily. In scenic Monterey County, California, many landowners have given voluntary easements covering thousands of acres, including parts of the spectacular shore line at Big Sur. In return for this landscape preservation, donors are protected against rising assessments that would force them to subdivide or sell.

Measures similar to the California open-space law have been adopted by several other states, including New York, Massachusetts, Maryland, Connecticut, and Wisconsin. The last-named state, for example, has purchased scenic easements at very low cost to preserve countryside along the Great River Road down the Mississippi. In New York State a different kind of easement provides access rights for fishermen along many miles of privately owned trout streams.

There are other useful tools for land planning. Agricultural land, for example, can be purchased by a public agency and leased back to the former owner with the proviso that its pastoral character be maintained. Ottawa, the Canadian capital, has pioneered in the use of this technique. In order to control its growth pattern the National Capital Commission of Canada purchased a semicircular belt of farmland and open space to the south of Ottawa. The inner margin of this green belt is about six miles from the center of the city and the belt itself is two and one-half miles deep and embraces some 37,000 acres. Most of the original farming and open-space uses continue as before.

The city is bounded on the north by the Ottawa River, beyond which a wedge-shaped extension of Gatineau Park eventually will comprise 75,000 acres. Although some phases of Ottawa's park and open-space program date from the turn of the century, the greater part of the master plan was developed by the French designer, Jacques Greber, after World War II. Ottawa today is a metropolitan area that would be regarded as a model by Olmsted himself. About 300,000 people live next door to more than 100,000 acres of superb parkland and green space that provide a permanent corridor of natural beauty for the capital city of Canada.

As the result of a more recent master plan, the county which surrounds Phoenix, Arizona, has laid out, as a land bank for the future, a peripheral group of regional parks embracing some 75,000 acres.

Pioneer open-space legislation has been enacted in Connecticut, which encourages communities to lower taxes on open land, to buy land or purchase interest in it, and to lease back purchased land subject to restrictions—all backed up by financial help and by the power of condemnation. As in Massachusetts, community conservation commissions have been established and given broad powers to protect and enhance their environments.

Vital as parks and open spaces are, they alone will not save our urban areas. Inevitably, cities will continue to be predominately man-made, and urban conservation must include the artificial as well as the natural. The most beautiful American cities are notable not only for their natural landscapes but also for the design and organization of the buildings and projects that make up the total environment. San Francisco, known for its superb natural setting, has enhanced its hills and bay with some of the most inspiring bridges constructed since the Romans demonstrated that bridge-building was an art as well as a science. Wisconsin's noted native son, Frank Lloyd Wright, argued throughout a long lifetime that buildings should blend with landscapes, and not vice versa; and Madison, his state's capital city, located among beautiful lakes and moraine hills, has made the most of its unusual terrain. Were he alive, Thoreau would have equal admiration for the 1,200-acre arboretum on Madison's outskirts, and a recently acquired 700-acre marsh within its city limits.

Such cities as Santa Fe, New Orleans, and Boston have been set apart by the distinctive way in which their historical sections have been preserved; and the renewal of downtown Philadelphia is an exciting example of both urban rebuilding and the conservation of historical landmarks.

Thanks to L'Enfant's grand design of 1791, Washington is one of the few large American cities planned from its inception. It also is one of the few large cities that has avoided skyscraper blight; in order that the Capitol, the Washington Monument, and other federal buildings should not be overwhelmed by taller structures, the height of buildings has been rigidly controlled. As a result, Washington has retained a green and spacious appearance, and its physical scale puts a premium on human values. The city's skyline and its vistas are dominated by the domes, monuments, and cathedrals that declare the aspirations of the American people.

City planning should put people first. Autos, freeways, airports, and buildings should not be allowed to dominate a city; each must take its own place in a balanced environment along with trees and parks, playgrounds and fountains. Just as there are certain areas from which skyscrapers should be excluded, so there should be more places where the automobile is off limits. Well-placed malls, plazas, promenades, and gardens can become oases inviting delight and giving a sense of order to living.

The crowding of our urban regions has caused us to look with new interest on such "useless" natural areas as marshes and swamps. A few years ago the suggestion that swamps might help save our cities would not have been taken seriously; but residents of Washington, D.C., Philadelphia, Madison, and some New Jersey communities have established large areas of swamp and marshland as permanent nature sanctuaries. It takes a perceptive eye to see the miracles of life in the woodlands and bogs where our forefathers would have seen only another opportunity to subjugate nature.

In all phases of city development we need to give free rein to imaginative designers like Eero Saarinen, whose Dulles International Airport near Washington is attracting worldwide acclaim; or Nathaniel Ownings, the main architect of San Francisco's glass-clad Crown-Zellerbach Building, located in its own park in downtown San Francisco. Public buildings, which are too often the scenes of shallow triumphs of penny-pinching officials, should set the pace in architectural design, in landscaping, and in the use of painting and sculpture.

From the standpoint of urban design, the size of many of our large cities has already reached the point of diminishing livability. Just as there is an optimum density of population within a given area, so there must be an optimum physical size for cities.

Long before universal double-decking and the overuse of vertical space make congestion intolerable, we must give more attention to the only practical alternative: the creation of new cities. The best of our "industrial parks" may point the way. Tax allowances and other incentives now encourage industries to locate in new areas in accordance with a master plan for land use.

Borrowing from the Ottawa pattern, some planners have visualized large urban constellations involving industrial parks, clustered housing, plentiful recreation areas and extensive green belts. These new cities should have a community life of their own and become creative centers of commerce and culture, "park cities" that would give priority to community living.

Innovations in technology are sure to provide opportunities for new kinds of urban planning: the development of nuclear reactors as a safe, cheap source of power; advances in air transportation; and the perfection of high-voltage, long-distance power transmission lines which will enable us to transmit electrical energy economically anywhere in the country will all aid the planners of tomorrow's cities. Together, these techniques will reverse the age-old process of locating cities only near waterways or along main arteries of commerce. Many planners are convinced that the principal hope for accommodating a much larger population in this country without impossible crowding lies in the development of new cities which will range in population from 30,000 to 300,000 people.

E. M. Forster once offered his countrymen words we might heed:

> If you desire to save the countryside there is only one way: through good laws rightly applied. . . . That is your only hope. A little has already been done: much more can be done in the future. It needs men of good will who can continue and work together lest destruction spread and cover the fields and the hills with its senseless squalor. Now is the moment. Soon it will be too late.

There is an unmistakable note of urgency in the quiet crisis of American cities. We must act decisively—and soon —if we are to assert the people's right to clean air and water, to open space, to well-designed urban areas, to mental and physical health. In every part of the nation we need men and women who will fight *for* man-made masterpieces and against senseless squalor and urban decay.

Like the mythical Antaeus, who was invincible as long as he was able to touch the earth, the urban American, if he keeps alive a saving reverence for the land, may accomplish a work of social engineering that will encourage the full participation of the best designers and artists and scientists and enlightened men of business in the building and re-creation of cities in which the finest human instincts can flourish.

CHAPTER XIII

Conservation and the Future

> *Conservation . . . can be defined as the wise use of our natural environment: it is, in the final analysis, the highest form of national thrift—the prevention of waste and despoilment while preserving, improving and renewing the quality and usefulness of all our resources.*
>
> —PRESIDENT JOHN F. KENNEDY
> Conservation Message to Congress (1962)

IF THE FORESTER and reclamation engineer symbolized the conservation effort during Theodore Roosevelt's time, and the TVA planner and the CCC tree planter typified the land program of the New Deal, the swift ascendancy of technology has made the scientist the surest conservation symbol of the 60's. His instruments are the atom-smasher, the computer, and the rocket—tools that have opened the door to an ultimate storehouse of energy and may yet reveal the secrets of the stars.

TR partially stopped the waste of resources, and his pilot conservation programs were a solid success. A generation later, Franklin D. Roosevelt's new agencies set out to rebuild the land, and the dams and development projects he instituted became the keystone of the conservation effort.

173

However, when the cadence of history accelerated with the onset of World War II, resource problems were either downgraded or transformed. FDR marshaled a maximum science-industry effort to produce the goods and weapons needed to win a global war. The supreme conservation achievement of this century, the fashioning of an almost self-renewing source of energy by the atomic scientists, was a direct result of the war effort. The first dramatic testing of an atomic pile at Chicago, in 1942, introduced the new role of scientists as the midwives of conservation. A hundred years of resource history were telescoped. The atomic physicists who uncovered the edge of an infinite dynamo brought fire, like the gods of Greek mythology, from seemingly inert elements, and allayed our fears of fuel shortage once and for all.

The surge of science was a boon to many areas of resource conservation. During the war years and those that followed, the alchemies of research brought new metals such as beryllium, germanium, columbium, molybdenum, and titanium into use, created synthetics and substitute products, and increased the usefulness of many raw materials. Investments in basic research in agriculture paid off with discoveries that gave wider insights into agronomy, while plant genetics and plant pathology yielded new strains, which, with new fertilizers and pesticides, made the granaries of American farmers overflow and gave us an opportunity to share our surplus and our science with other nations. Similarly, research in animal husbandry increased our ability to raise livestock, and science encouraged the owners of public and private forest lands to apply the tree-farming techniques fostered by the successors of Gifford Pinchot.

These triumphs of technology have, in the 1960's, lent a note of optimism to the reports of most resource experts.

Today, we are told, technology carries in its hands the keys to a kingdom of abundance, and sound solutions to many conservation problems rest largely on adequate research and efficient management. At last, long-range resource planning is becoming an indispensable aide to science in assuring an abundance of resources for human use.

Ironically, however, these very successes of science have presented a new set of problems that constitute the quiet crisis of conservation in this decade. It began with the inrush to the cities at the outset of World War II, and intensified with each new advance of technology. Our accomplishments in minerals and energy, in electronics and aircraft, in autos and agriculture have lifted us to new heights of affluence, but in the process we have lost ground in the attempt to provide a habitat that will, each day, renew the meaning of the human enterprise. A lopsided performance has allowed us to exercise dominion over the atom and to invade outer space, but we have sadly neglected the inner space that is our home. We can produce a wide range of goods and machines, but our manipulations have multiplied waste products that befoul the land, and have introduced frightening new forms of erosion that diminish the quality of indispensable resources and even imperil human health. The hazards appear on every hand; many new machines and processes corrupt the very air and water; in what Rachel Carson has called "an age of poisons," an indiscriminate use of pesticides threatens both man and wildlife; and the omnipresent symbol of the age, the auto, in satisfying our incessant demand for greater mobility, has added to the congestion and unpleasantness of both cities and countrysides.

The conservation effort was confused and side tracked by the cataclysmic events that began in 1939. In the two decades that followed, public men were so preoccupied by

the urgent issues of the hot and cold wars that none tried,
as Pinchot and the two Roosevelts had done, to expand the
conservation concept and apply it to the new world of
natural resources and the new problems of land stewardship.

As a result of this failure to keep the conservation idea
abreast of the times, such successful conservers as the scientific industrialists and scientific farmers seldom consider
themselves conservationists at all, while many modern disciples of Thoreau and Muir have narrowed their concern to
park, forest, wildlife, or wilderness problems.

With the passing of each year neglect has piled new
problems on the nation's doorstep. Some brilliant successes
—in electronics, atom physics, aerodynamics, and chemistry, for example—encouraged a false sense of well-being,
for our massive ability to overpower the natural world has
also multiplied immeasurably our capacity to diminish the
quality of the total environment. Our water husbandry
methods have typified these failures. At the same time that
our requirements for fresh water were doubling, our national sloth more than doubled our water pollution. We
now are faced with the need to build 10,000 treatment
plants and to spend $6,000,000,000 to conserve water
supplies.

Much of our river development proceeded on an uncoordinated basis (although regional planning, which Major
Powell would have applauded, had an inning when the
Upper Colorado River Storage Project was passed in 1956).
The incursions of industry, agriculture, and urbanization
into the marshlands were reducing waterfowl populations.
Acreage in new parklands created by federal and state governments was too sparse to be significant: in 1940, 130,-
000,000 Americans had a spacious National Park system
of 22,000,000 acres; twenty years later, a population
which had grown to a more mobile 183,000,000 inherited

an overcrowded system that had been enlarged by only a few acres. Of 21,000 miles of ocean shore line in the contiguous 48 states, only 7 per cent was reserved for public recreation.

In addition, the most eroded lands in the United States —the overused grasslands of the Western public domain— were not restored to full fertility despite the new American awareness of the importance of soil conservation. Asphalt inroads of city subdividers in search of quick profits were so ill-conceived that stream valleys and open space were obliterated.

In the postwar period, unfortunately, most Americans took their out-of-doors for granted. It was a fact that pressures were growing each year to despoil our few remaining wilderness areas; Americans who were accustomed to outdoor recreation as a way of life—with access to public areas for hunting, fishing, hiking, and swimming—found overcrowding increasing each year. Most state and city governments faced so many growth problems that they had little time for foresight in planning their over-all environment. It was a sad fact, also, that the men, women, and children of America the Beautiful became the litter champions of the world. Each year about 5,000,000 battered autos are added to our junk yards. Aided by industries that produce an incredible array of boxes, bottles, cans, gadgets, gewgaws, and a thousand varieties of paper products, our landscape litter has reached such proportions that in another generation a trash pile or piece of junk will be within a stone's throw of any person standing anywhere on the American land mass. Our irreverent attitudes toward the land and our contempt for the Indians' stewardship concepts are nowhere more clearly revealed than in our penchant to pollute and litter and contaminate and blight once-attractive landscapes.

The promised land of thousands of new products, ma-

chines, and services has misled us—and the conservation
movement, which should have become an intricate and
interlocking effort on a hundred fronts, was itself disorgan-
ized and outdated.

Simultaneously, the steep upsurge of population and the
pell-mell rush to enlarge our cities changed our people-to-
people ratio and our attitudes toward the land with it. Indifference to the land was also accelerated by new seductions
of spectatorship, the requirements of industrial growth, and
air-conditioned advantages that made glassed-in living so
appealing. The predictable result was that sedentary, city-
bound citizens were encouraged to acquiesce in the diminu-
tion of the spaciousness and freshness and green splendor
of the American earth.

Intoxicated with the power to manipulate nature, some
misguided men have produced a rationale to replace the
Myth of Superabundance. It might be called the Myth of
Scientific Supremacy, for it rests on the rationalization that
the scientists can fix everything tomorrow.

The modern land raiders, like the public-land raiders of
another era, are ready to justify short-term gains by seeking to minimize the long-term losses. "Present the repair
bill to the next generation" has always been their unspoken
slogan.

As George Perkins Marsh pointed out a century ago,
greed and shortsightedness are conservation's mortal enemies. In the years ahead, the front line of conservation will
extend from minerals to mallards, from salmon to soils,
from wilderness to water, but most of our major problems
will not be resolved unless the resource interrelationships
are evaluated with an eye on long-term gains and long-
term values.

Large-scale conservation work can no longer be accomplished by the flourish of a president's pen or through
funds appropriated to fight a business depression. If we are

to preserve both the beauty and the bounty of the American earth, it will take thoughtful planning and a day-in and day-out effort by business, by government, and by the voluntary organizations.

If the area of individual involvement is enlarged, if enough modern Muirs step forward to fight for "legislative interference" to save land and check its despoilment, the conservation movement can become a sustained, systematic effort both to produce and to preserve.

Full-fledged collaboration of science and industry and government, quickened by the spur of business competition, will enable us to write bright new chapters in the conservation of some resources. The continuing revolution in research should give us the means to harness the tides of Passamaquoddy, interconnect the electric-power systems of whole regions, economically extract fresh water from the seas, turn vast oil shale beds into oil, and, by discovering the innermost secrets of fission and fusion, allow us to "breed" energy from rocks.

Government leadership and government investment, however, must continue to play the larger role in traditional conservation work. In a matter of decades many regions will confront an insistent water crisis. Water conservation must always be primarily a public endeavor. It is already plain that regional planning, basin-wide water regimens, transmountain diversions of water from areas of surplus to more arid watersheds, sustained yield management of underground aquifers, and the development of techniques for pollution control and the re-use and recycling of water will be needed to save the day for water-short areas of the United States. To achieve this we must begin now to train a fully adequate corps of hydroscientists and to develop an awareness of the vital elements of a water-conservation program.

In addition, the forested land of our country must be

managed more intensively to achieve much higher wood yields, and public forest lands must have much wider use if we are to provide adequate outdoor opportunities for our citizens. Likewise, our efforts to save soil, to control stream pollution, and to repair the land damage of the past must be enlarged and intensified.

The quiet conservation crisis is the end product of many forces. Its threat is all the more serious because most harm involves subtle erosion and contamination, and because motives of commercial profit do not enlist public support. Quick action can be expected only when threats to the public health or public convenience are imminent. The larger task will not be undertaken unless a quickening conscience brings us to act now to protect the land for future generations.

President Kennedy's preservation-of-environment program is a response to the quiet crisis, and it points the way toward the main arenas of conservation action in the years ahead. It concerns wilderness and wildlife and parklands and the whole spectrum of outdoor resources. The American out-of-doors was studied with thoroughness and vision from 1959 to 1962 by the Outdoor Recreation Resources Review Commission. Its report is a landmark analysis of our past failures and present opportunities in the use and protection of our environment.

As inheritors of a spacious, virgin continent we have had strong roots in the soil and a tradition that should give us special understanding of the mystique of people and land. It is our relationship with the American earth that is being altered by the quiet crisis, our birthright of fresh landscapes and far horizons. Unless we are to betray our heritage consciously, we must make an all-out effort now to acquire the public lands which present and future generations need. Only prompt action will save prime park, forest, and shore line and other recreation lands before they are pre-empted for other uses or priced beyond the public purse.

The Land and Water Conservation Fund proposed by President Kennedy may mark a turning point in conservation history. If the states are to provide leadership before it is too late, if the few remaining spacious seashores are to be preserved for all of the people, if wildlife values are to be permanently protected and our National Park, Forest, and Wildlife Refuge systems are to be rounded out by the addition of the remaining suitable lands, the task must begin immediately and be completed within the next three decades.

The status we give our wilderness and near-wilderness areas will also measure the degree of our reverence for the land. American pioneering in establishing National Parks and in promoting the wilderness concept is already being emulated in many parts of the world today. Many nations no longer have the option of preserving part of their land in its pristine condition. We must take ours up before it is too late. A wilderness system will offer man what many consider the supreme human experience. It will also provide watershed protection, a near-perfect wildlife habitat, and an unmatched science laboratory where we can measure the world in its natural balance against the world in its man-made imbalance. The case for wilderness protection was stated eloquently for all time by Senator Clinton P. Anderson of New Mexico when he said:

> Wilderness is an anchor to windward. Knowing it is there, we can also know that we are still a rich Nation, tending to our resources as we should—not a people in despair searching every last nook and cranny of our land for a board of lumber, a barrel of oil, a blade of grass, or a tank of water.

In Alaska we have a magnificent opportunity to show more respect for wilderness and wildlife values than did our forebears. The wonders of the wilderness still abound there; if we spoil them, we cannot excuse their defilement with pleas of ignorance.

We have an opportunity, too, to take a leaf from FDR's book and establish a permanent Conservation Corps to rehabilitate and renew our public lands. Generations to follow will judge us by our success in preserving in their natural state certain rivers having superior outdoor recreation values. The Allagash of Maine, the Suwannee of Georgia and Florida, the Rogue of Oregon, the Salmon of Idaho, the Buffalo of Arkansas, and the Ozark Mountain rivers in the State of Missouri are some of the waterways that should be kept as clean, wild rivers—a part of a rich outdoor heritage. We must act to provide a habitat in which the fish and wildlife sharing our planet may thrive. Our environment-preservation work will lack balance unless our highway building includes a national system of scenic roads.

The quiet crisis demands a rethinking of land attitudes, deeper involvement by leaders of business and government, and methods of making conservation decisions which put a premium on foresight. With the acumen of our scientists we can achieve optimum development of resources that will let us pluck the fruits of science without harming the tree of life. Once we decide that our surroundings need not always be subordinated to payrolls and profits based on short-term considerations, there is hope that we can both reap the bounty of the land and preserve an inspiriting environment.

The Muirs and Olmsteds and Pinchots of the decades ahead will surely fail unless both our business and public budgets embrace conservation values. Enlightened leaders of the business community are already pointing the way: such companies as Lever Brothers, Johnson's Wax, and the Connecticut General Life Insurance Company have demonstrated that a beautifully designed building is the most attractive form of advertising. Conservation will make headway when it is patently good business for companies to invest in programs of education and practices of production which emphasize both conservation and industrial efficiency. Conservation statesmen must prove that profits and the conservation cause are compatible if we are to succeed in making an attractive and orderly environment part of our national purpose. At present, many of our policies actively conspire against conservation, and the conservation-minded businessman too often finds himself at an intolerable competitive disadvantage if he implements his convictions.

The proper control of waste products or polluting materials, or the reclamation of strip-mine areas, cost money. For any one state or any one region to allow its enterprisers an economic advantage by permitting damage illegal elsewhere is a repetition of the nineteenth-century story of the forest raiders and hydraulic miners. Our air and water resources are essential to life. Water and the air masses are in constant motion, and it will take uniform laws, national in scope, to put competing industries on an equal footing. Where our laws make land reclamation and pollution abatement a normal part of the cost of doing business, enlightened businessmen are already working with the conservation cause. Once intolerable competitive advantages are eliminated, researchers will quickly devise machines and gadgets to minimize the damage. But environment restoration and preservation can succeed only if we pay as we go.

We have reached the point in our history where it is absolutely essential that all resources, and all alternative plans for their use and development, be evaluated comprehensively by those who make the over-all decisions. As our land base shrinks, it is inevitable that incompatible plans involving factories, mines, fish, dams, parks, highways and wildlife, and other uses and values will increasingly collide. Those who decide must consider immediate needs, compute the values of competing proposals, and keep distance in their eyes as well. For example, technical innovations which will have a widespread effect on other resources and on living values shared by all must be assessed in advance. Chemical contamination, the disposal of radioactive wastes, and sonic boom are examples of present or coming problems which will require the careful measurement of social costs against social benefits.

Our mastery over our environment is now so great that the conservation of a region, a metropolitan area, or a valley is more important, in most cases, than the conservation of any single resource. Complex decisions will require sophisticated judgments that weigh all elements and explore all possible alternatives. Slum valleys and regional slums will be the result, unless we put our resources to their highest and best use.

As the area of conflict and overlap increases we must constantly improve our decision-making techniques. Nor must we be afraid to decide the toughest of issues: practices that defer necessary decisions can also be a threat to the national welfare.

Geography has aways been a global science and conservation must now become a truly global concept if the optimum use of resources is to be achieved. Nature's rules still obtain, and all parts of the natural world, from minerals and marine life to the gulf streams of the ocean and jet streams of the upper atmosphere, obey a single set of laws.

It is the seven seas themselves, the one remaining largely unspoiled, untapped resource, which now represent the largest remaining frontier of conservation on this earth.

The atmosphere and the oceans are the two resources that are owned by all of the people of the world. Yet, save for a few farsighted treaties, we have no plan of management for these common resources, and oceanographers are still at the outer edge of the secrets of the sea. The ocean domain includes the submerged 71 per cent of the earth's surface, and in its depths are an immense reserve of mineral and marine resources for the future: for example, once the enormous schools of "trash" fish are converted into edible food or fish flour, the protein diet deficiency afflicting two-thirds of the people of the world can be partially alleviated.

Inadequate research into marine resources and the absence of international planning have meant that some resources are overexploited while others remain underdeveloped. With the exception of a few notable agreements the law of hunt-and-kill is still the code of the sea. Only timely international conservation agreements will avert the spectacle of a resource raid to dwarf those of the past on the fur seal and the sea otter. The oceans can be the most fruitful field for international co-operation in conservation, if the nations will turn in time to the principles of sustained-yield management.

The one factor certain to complicate all of our conservation problems is the ineluctable pressure of expanding population. Our resource planners operate in a bureaucratic trance, assuming that the population of the United States will inevitably double by the year 2000. An all-too-common corollary assumption is that life in general—and the good, the true, and the beautiful in particular—will somehow be enhanced at the same time. We have growth room in this country, but the time has come for thoughtful men and women to ask some basic questions about our land-people equation. Our whole history demonstrates that his physical environment has an enormous influence upon man. Are not such inquiries as these, then, pertinent to the future course of human enterprise: What is the ideal "ecology of man," the ideal relationship of the human population to environment? Is man subject to the laws of nature, which hold that every species in any environment has an optimum population? How much living space do human beings need in order to function with maximum efficiency and to enjoy maximum happiness?

It is obvious that the best qualities in man must atrophy in a standing-room-only environment. Therefore, if the fulfillment of the individual is our ultimate goal, we must soon determine the proper man-land ratio for our continent.

Our future will be linked increasingly with the success of other peoples in dealing with their resources. The American economy already consumes over 30 per cent of the world's raw material production, and resource interdependence among nations increases each year. The Peace Corps program and the second phase of our foreign aid effort largely involve the export of conservation know-how. That foreign students already outnumber Americans in the mining courses offered by our graduate schools serves to measure the extent of our educational contributions. We are generous enough and practical enough to share our resource insights with the farmer in Pakistan, the Peruvian fisherman, and the game warden of east Africa. In return, we ought to accept gladly the land lessons they can teach us. The region of the river valley is now the proper setting for resource planning, but technology will soon make international conservation planning a necessity. The treaty that

made the Antarctic a scientific preserve, and the world-
wide co-operation evinced by the International Geophysi-
cal Year, are signposts of hope for the future.

In the years ahead, nations can either compete ruthlessly
for resources, in a context of scarcity, or co-operate, respect
the laws of nature, and share its abundance. Resource interdependence and the common management of those resources owned in common will enlarge the area of unified
action and do much to encourage world order. The growth
of a world-wide conservation movement might be a gyro-
scopic force in the world politics. The most influential
countries of the future surely will be those that bring
desalted water to arid lands and use their scientific discoveries to advance the welfare of all mankind.

Internationally, we need new forms of co-operation in
order to realize the full potential of all natural resources.
Domestically, we must have a ground swell of concern over
the quiet crisis, which could culminate in a third wave of
the conservation movement.

The creation of a life-giving environment can go hand-
in-hand with material progress and higher standards of
husbandry, if, in President Kennedy's words, we make time
"our friend and not our adversary."

CHAPTER XIV

Notes on a Land Ethic for Tomorrow

We abuse land because we regard it as a commodity belonging to us. When we see land as a community to which we belong, we may begin to use it with love and respect.

—Aldo Leopold
"A Sand County Almanac"

BEYOND all plans and programs, true conservation is ultimately something of the mind—an ideal of men who cherish their past and believe in their future. Our civilization will be measured by its fidelity to this ideal as surely as by its art and poetry and system of justice. In our perpetual search for abundance, beauty, and order we manifest both our love for the land and our sense of responsibility toward future generations.

Most Americans find it difficult to conceive a land ethic for tomorrow. The pastoral American of a century ago, whose conservation insights were undeveloped, has been succeeded by the asphalt American of the 1960's, who is shortsighted in other ways. Our sense of stewardship is uncertain partly because too many of us lack roots in the soil and the respect for resources that goes with such roots.

Too many of us have mistaken material ease and comfort for the good life. Our growing dependence on machines has tended to mechanize our response to the world around us and has blunted our appreciation of the higher values.

There are many uprooting forces at work in our society. We are now a nomadic people, and our new-found mobility has deprived us of a sense of belonging to a particular place. Millions of Americans have no tie to the "natural habitat" that is their home. Yet the understanding of the grandeur and simplicity of the good earth is the umbilical cord that should never be cut. If the slow swing of the seasons has lost its magic for some of us, we are all diminished. If others have lost the path to the wellsprings of self-renewal, we are all the losers.

Modern life is confused by the growing imbalance between the works of man and the works of nature. Yesterday a neighbor was someone who lived next door; today technology has obliterated old boundaries and our lives overlap and impinge in myriad ways. Thousands of men who affect the way we live will always remain strangers. An aircraft overhead or an act of air or water pollution miles away, can impair an environment that thousands must share. If we are to formulate an appropriate land conscience, we must redefine the meaning of "neighbor" and find new bonds of loyalty to the land.

One of the paradoxes of American society is that while our economic standard of living has become the envy of the world, our environmental standard has steadily declined. We are better housed, better nourished, and better entertained, but we are not better prepared to inherit the earth or to carry on the pursuit of happiness.

A century ago we were a land-conscious, outdoor people: the American face was weather-beaten, our skills were muscular, and each family drew sustenance directly from the land. Now marvelous machines make our lives easier, but we are falling prey to the weaknesses of an indoor nation and the flabbiness of a sedentary society.

A land ethic for tomorrow should be as honest as Thoreau's *Walden*, and as comprehensive as the sensitive science of ecology. It should stress the oneness of our resources and the live-and-help-live logic of the great chain of life. If, in our haste to "progress," the economics of ecology are disregarded by citizens and policy makers alike, the result will be an ugly America. We cannot afford an America where expedience tramples upon esthetics and development decisions are made with an eye only on the present.

Henry Thoreau would scoff at the notion that the Gross National Product should be the chief index to the state of the nation, or that automobile sales or figures on consumer consumption reveal anything significant about the authentic art of living. He would surely assert that a clean landscape is as important as a freeway, he would deplore every planless conquest of the countryside, and he would remind his countrymen that a glimpse of grouse can be more inspiring than a Hollywood spectacular or color television. To those who complain of the complexity of modern life, he might reply, "If you want inner peace find it in solitude, not speed, and if you would find yourself, look to the land from which you came and to which you go."

We can have abundance and an unspoiled environment if we are willing to pay the price. We must develop a land conscience that will inspire those daily acts of stewardship which will make America a more pleasant and more productive land. If enough people care enough about their continent to join in the fight for a balanced conservation program, this generation can proudly put its signature on the land. But this signature will not be meaningful unless we develop a land ethic. Only an ever-widening concept and higher ideal of conservation will enlist our finest impulses and move us to make the earth a better home both for ourselves and for those as yet unborn.

PART TWO

THE NEXT GENERATION

CHAPTER XV

Ecology to the Forefront:

RACHEL CARSON AND *SILENT SPRING*

A scientist in the grand literary style of Galileo and Buffon, she has used her scientific knowledge and moral feeling to deepen our consciousness of living nature and to alert us to the calamitous possibility that our short-sighted technological conquests might destroy the very sources of our being.
—CITATION OF THE AMERICAN ACADEMY OF ARTS AND LETTERS (1963)

W HEN *Silent Spring* appeared in the summer of 1962, it caused a furor in scientific and business circles that spilled over into the highest levels of the federal government. It challenged actions and assumptions of the powerful chemical and agribusiness industries that exemplified the prevailing superoptimism about the transforming power of the "new science." And it spurred new lines of thought about resources and the limits of technology that began to alter the thinking of my generation.

This combative challenge startled the smug leaders of the scientific community. While it was foreseeable that the gauntlet would be thrown down by a biologist schooled in the conservation principles popularized by the two Roosevelt presidents, few would have surmised that the author of such a controversial book would be a woman scientist who specialized in marine zoology and worked for the federal government.

A shy, bookish person, Miss Carson grew up on a farm in the lower valley of the Allegheny River, won academic honors at Pennsylvania College for Women, and later acquired an M.A. in zoology from Johns Hopkins University. Miss Carson was a part-time instructor in biology at Maryland University when she passed a civil service exam in 1936 and went to work as an aquatic biologist for one of Washington's old-line conservation agencies, the U.S. Bureau of Fisheries. Once her exceptional writing skills were discovered, she spent most of her sixteen government years in the Fish and Wildlife Service as a science writer preparing pamphlets and press releases to educate the public about the importance of preserving life-giving habitats for fish and wildlife.

Like Charles Darwin, Rachel's insatiable curiosity about the functioning of nature's systems lifted her out of a bureaucratic groove into a realm where she could expand human knowledge about the earth. She was much more than a fine writer. What made her one of this century's most influential scientists was a perfectionism that caused her to dig out and digest the original work of other scientists before she formed opinions and began composing the powerful, poetic writing that made her an exceptional nature writer and a best-selling author.

Two decades before *Silent Spring*, praise generated by an article Rachel Carson wrote about undersea life for the *Atlantic Monthly* led to her first book, *Under the Sea Wind*, and persuaded her that she knew how to explain nature's mysteries and make science an exciting subject for millions of nontechnical readers. That confidence encouraged her to devote all her spare time after World War II to a regimen of research and writing that, years later, resulted in a prestigious international best-seller, *The Sea Around Us.* The success of her books was testimony that science could be a popular topic.

A wide-ranging profile of the oceans, *The Sea Around Us* was an immense undertaking for a full-time government employee. Carson's subject was a planet, and in an exhaustive search for precise data she corresponded with many of the world's eminent oceanographers to gain insights and test the accuracy of her conclusions. It was her tireless determination to ferret out every pertinent fact—and to understand and explain the interactions of entire living communities—that made each of her writing projects a time-consuming enterprise.

Ultimately this holistic approach, repeated and refined in her 1955 sequel, *The Edge of the Sea,* became the springboard which positioned her to write a final, pioneering volume that catapulted the concept of ecology into millions of minds—and into the global dialogue. Miss Carson had schooled herself to see the web of life in a Maine tide pool, and she wanted to share with others the precious knowledge she had garnered about nature's life-sustaining systems. But Rachel Carson wanted to do more than popularize science. In one of her rare public addresses she said, "The materials of science are the materials of life itself . . . It is impossible to understand man without understanding his environment and the forces that have molded him physically and mentally."

Miss Carson began writing *Silent Spring* in 1958 when her friends in Massachusetts and on Long Island, whose property had been drenched with DDT, persuaded her to consider composing a protest article for a national magazine. She did so, but her proposal was rejected by *Reader's Digest* and other major magazines. Subsequently she toyed with the idea of collaborating with other writers to produce a short book on the subject of pesticides, but soon realized that if a cogent book was to be written, she would have to write it herself.

Rachel Carson's initial survey revealed that the problem was global, not local: toxic chemicals even more lethal than DDT were already in the food chains and were presenting a peril to the future of all living things. Once she saw the enormity of the pesticide issue, she was ready to devote all of her energies on what she began to view as a crusade. By late fall she was "working like mad" digesting the growing pile of documents sent to her by scientists from all parts of the world.

From the outset Miss Carson knew that she was on a collision course with powerful adversaries. Like penicillin, DDT was viewed as a boon to humankind. It had shielded U.S. soldiers from diseases in World War II, public health doctors on all continents were prematurely giving it credit for wiping out typhus and malaria, and the world's agricultural experts were attributing dramatic increases in food output to DDT and its deadly cousin-poisons.

In those days the chemical industry was marching under the advertising slogan "Better Living Through Chemistry," and to the public DDT symbolized the industry's contributions to human betterment. To criticize DDT was to attack a sacred cow, for it bore the seal of approval of the National Academy of Science, and Paul Mueller of Switzerland, the chemist who discovered it, had won a Nobel Prize for his work. Inevitably, DDT's prowess as an exterminator inspired products of greater potency, and one of the grim ironies of *Silent Spring* is that even as Rachel was marshaling her facts in 1960, biochemists in an industrial laboratory were perfecting the superpoison, Aldicarb, which would kill more than three thousand people at Bhopal, India, in the 1980's.

Because she was writing a polemical book opposing a powerful industry, Carson suspected that the chemical barons would mount a counterattack which might propel her into an ugly national scientific dispute. Although her combative instincts had never been tested in this way, and one friend warned her that her new work would not be a best-seller "no matter how beautifully it is written," she plunged ahead, determined to take whatever time was necessary to build her argument on an unshakable foundation of facts.

The decision to present *Silent Spring* in an ecological context grew, in part, out of Carson's conviction that Atomic Age science had lost sight of the vital functions performed by nature's biological systems. She would begin by describing how our lives are interlinked with these systems, outline what the new poisons were doing to disrupt the environment, and end with a plea for restraint and a program of action based on ecological balance.

Rachel was dismayed by the arrogance of specialists who exhibited little interest in following the biological side effects of their efforts to control nature. As the bearer of ecology's banner, she demanded that the scientific community weigh all its activities in relation to the total environment. A pioneer advocate of the public's "right to know," Miss Carson rejected the idea that technical subjects should be left to technicians. She expressed the opinion in 1952 that it was a mistake to allow public policies to be made by ". . . a small number of human beings, isolated and priestlike in their laboratories."

Rachel Carson realized early that, unlike her other books, *Silent Spring* had to be a scientific brief on behalf of a cause. As her research progressed, she knew that she had to relate the facts about the new "age of poisons" to everyday life. She resolved this problem by using case studies of unbridled spraying to translate ecological truths into concepts the average reader could comprehend and identify with.

The first eleven chapters of *Silent Spring* constitute a continental tour which could have been subtitled, "The Toxic Chemical Follies of the 1950's." Keeping a steady focus on nature's "web of interwoven lives"—and on the human consequences of each spraying foray—Rachel Carson described the outcome of efforts by "control men" to eradicate insects and plants in a huge area of North America that included a Nova Scotia forest, Detroit suburbs, Illinois farms, the Yellowstone River, rice fields in California, cotton plantations in Mississippi, and hillsides of apple orchards in Virginia.

Rachel suspected her findings would be questioned by most of her readers, so she supplied them with local facts demonstrating the step-by-step process by which supposedly beneficent superpoisons were killing their fish and wildlife—and invading the bodies of their children. What made *Silent Spring* an ecology primer for millions was Carson's skill as a simplifier, her uncanny ability to use science as a mirror, reflecting its subtleties in the materials of life itself.

Making *Silent Spring* a book about the spacious concepts of ecology was a masterstroke in more ways than one. It shifted the debate over pesticides into a context where ecological, not economic, values would predominate. And it enabled the author, who deplored the tunnel vision of specialized science, to force the chemical industry's scientists out of their bunkers and into a contest on the terrain of the total environment.

Miss Carson's discussion of the health hazards posed by pesticides that were accumulating in the fatty tissues of humans added a vital dimension to her book. Although laboratory tests had shown that some of the new pesticides were carcinogens, the prestigious physicians she drew into her circle cautioned that more research would be needed before cancer-causing agents could be identified and linked to exposures to specific carcinogens. Her three chapters about pesticides and cancer summarized the findings of researchers who were trying to understand the mutations that cause human cells and chromosomes to malfunction and produce life-destroying tumors. Miss Carson knew she was out of her field, so she was careful to let her experts and their facts speak for themselves.

But she was appalled by evidence that pointed to a link between the proliferation of radiation and toxic chemicals and increases in the incidence of cancer in children, and she concluded her book by posing questions about environmental protection that, to this day, haunt those who formulate our public health policies. If DDT and other toxic chemicals are carcinogens, she asked, wasn't a "safe dose" a zero dose? If children are most vulnerable to cancer-causing poisons, shouldn't legal barriers be devised to prevent carcinogens from entering food chains? And wasn't "the human price" too high for society to allow an industry to fill ". . . the environment with chemicals that have the power to strike directly at the chromosomes"?

Rachel Carson had a pragmatic attitude toward modern life. She refused to accept the proposition that damage to nature's system was an inevitable feature of "progress." She wanted an "ecology for man" to counterbalance the excesses spawned by Atomic Age arrogance. As a scientist, she did not call for a ban on all chemical insecticides; rather, she favored biological controls and concluded her text with a review of some promising alternatives offered by entomologists who were exploring "the other road" to pest control.

Miss Carson believed her book would shake the rafters of the scientific establishment. She also knew that her industrial adversaries would try to demolish her arguments with unrelenting counterattacks by prestigious consultants who worked for them. To meet these exigencies, the author attached a fifty-five-page appendix of "principal sources" listing over 600 of the thousands of documents she had digested during her four years of research. This appendix was Rachel Carson's way of saying to her critics: "Here is your substantiation. Tear it apart if you can."

Miss Carson's core argument gained credence when President John F. Kennedy's Science Advisory Committee supported it, but there are still debates about *Silent Spring*'s influence on the evolution of our thinking about earth's inhabitants and their resources. It would be difficult to understate Rachel Carson's contribution to science and culture. Her thinking had a worldwide impact, and we have yet to assess the reach of her legacy. Among the life scientists of this century she was supremely influential in helping humankind understand the changing relationships between the stewardship of earth's resources and the necessity of efforts to sustain and enhance life-promoting activities on this planet.

It is undeniable that *Silent Spring* had a wide reach in the 1960's. It inspired the Swedes to enact stringent ecological laws, and it encouraged other countries to begin fashioning effective environmental protection policies. The pace of environmental reform was slow in this country, but this did not surprise Rachel Carson. Her Washington experience told her it would be difficult to curb activities supported by powerful industries. However, she lived long enough to witness the beginning of action that would culminate in an environmental revolution. The summer *Silent Spring* appeared, the Department of Interior adopted a policy to discourage the use of pesticide poisons. More important, Carson's thesis inspired a commitment to what we called "the total environment approach" that permeated her old department and resulted in a yearbook, *Man ... An Endangered Species?,* which became a government best-seller two years after her death.

Silent Spring was, in effect, the first global environmental impact statement prepared by a scientist. *The Sea Around Us* had been translated into thirty-two languages, and the author had climbed to a vantage point where eminent scientists on all continents were eager to help her compose a work that would address environmental problems in a context that included all living things on earth.

It took U.S. lawmakers more than two years to pass pertinent laws, but it is undeniable that Rachel Carson's concepts inspired both the enactment of a National Environmental Policy Act (mandating that all government agencies assess environmental impacts before new projects are undertaken) and the eventual establishment of a federal Environmental Protection Agency to control all sources of pollution.

Silent Spring also served as an antidote to the technological optimism that was assuming the status of a U.S. religion in the 1960's. Miss Carson's insistence that the deleterious side effects be taken into account tempered Atomic Age ebullience and cast a shadow of skepticism over the assumption that technologists could come up with "quick fix" solutions if catastrophic mistakes were made.

I am convinced that Rachel Carson's most important accomplishment related not to the immediate action she aroused, but to a reorientation of human thought. Some have questioned her achievements by observing that *Silent Spring* did not stop the pesticide threat nor topple the walls of industrial Jerichos. But she raised humanity's awareness of earth's laws, altered our perception of our relationship to the living world, and began an educational process by which ecological precepts entered the common vocabulary.

Miss Carson did not make new discoveries, nor did she develop an original theory about the physical world. Rather, she was a person with prophetic gifts who synthesized the insights of other scientists. She was, in short, a different Darwin of a different century.

CHAPTER XVI

The Flowering of Environmental Activism:

DAVID BROWER AND THE RISE OF THE SIERRA CLUB

Thank God for David Brower; he is so extreme he makes the Izaak Walton League look reasonable.
—IZAAK WALTON LEAGUE MAGAZINE (1966)

ENVIRONMENTALISM did not appear suddenly on the national scene as a concept that magnified the meaning of conservation. Like a branch grafted onto a healthy tree, it began to flourish in the 1960's as the conservation movement widened its concerns and causes. The dynamics of that growth can be traced by following the career of David Brower, a modern John Muir who played a dominant role in transforming the parochial Sierra Club of California into a militant national champion of environmental reforms.

David Brower's life began in 1912 almost literally where John Muir's left off. Muir did not have a son to carry on his work, but had he dreamed on his deathbed in 1914 of a spiritual heir who would spread his cause across the continent, a Brower-like image might have floated through his mind.

Brower, like Muir, was a shy person who discovered his core convictions in the out-of-doors. He took extensive courses in Muir's "University of the Wilderness" and became a mountaineer. Muir was a saunterer who tuned his senses to nature's subtle rhythms. Brower was a rock climber who pioneered routes to the summits of thirty-three of the highest peaks of the Sierra Nevada range. Both were skilled conversationalists who developed a reverence for the printed word—Muir as a nature writer who added new rooms to the mansion constructed by Henry David Thoreau, Brower as conservation's most successful editor and publicist.

Yet as public men their careers took very different paths. John Muir, a reticent man who was never comfortable in public forums, influenced opinion mainly with his pen. His achievements as a lobbyist were modest. He lost his Hetch Hetchy fight, and many of the legislative triumphs associated with his name were the work of his adroit friend, Robert Underwood Johnson of *Century* magazine. Muir was essentially a private person, a writer who dabbled in the politics of legislative interference. Brower, by contrast, threw off the diffidence of his youth, developed a zest for the raw meat of controversy, and became a lion on the nation's pathways of political power. Muir, the quiet clubman, did not like to make speeches, whereas Brower became an evangelist who thrived on discourse and ranged across the country to bring his gospel of environmental awareness to the masses.

John Muir is a force today because his wilderness philosophy and his message about the fragile interdependence of all living things embody the central concepts of our ecological conscience. David Brower will be remembered in the twenty-first century as a great teacher and salesman who showed citizens in a burgeoning democracy how militant activism could preserve some of the earth's irreplaceable temples.

After John Muir died in 1914, his organization became little more than a regional hiking club, preoccupied with the resources in its own backyard. Even during the heyday of the New Deal, the club played second fiddle when Secretary Harold Ickes initiated action to create a King's Canyon National Park in the southern reaches of the "range of light" it treasured. When David Brower was hired as the first executive director, the Sierra Club's governing board was charting a new course.

Conservation was caught in an ebb tide in the 1950's. Opportunistic Western congressmen were conducting a campaign to "return" huge blocks of forest and grazing lands (given status as national lands by Teddy Roosevelt and his predecessors) to their respective states. President Eisenhower deemed the pollution that was choking American rivers a local problem. And, in an alliance with elements of the automobile industry, the National Park Service was concentrating on a program to build more roads and visitor facilities in the nation's parks.

It was a five-year fight over a plan developed by the Truman administration to build a dam at Echo Park in Dinosaur National Monument that rejuvenated the conservation movement by forcing it into a fight to stop this new Hetch Hetchy. As a result of rousing magazine articles written by historian Bernard De Voto, Echo Park became a battle cry for postwar conservationists. De Voto was an incisive, feisty interpreter of the history of Western development, so it was no accident that he was among those who picked up John Muir's fallen lance. Benny De Voto did not have Muir's eloquence, but his invigorating prose helped galvanize a new round of activism that encouraged Brower and other conservation leaders to shape national issues and carry them to Capitol Hill.

David Brower made his Washington debut at congressional hearings on the Bureau of Reclamation's Echo Park proposal. Sensing that an appeal based on esthetics would fall short, he conferred with experts who counselled him to question the basic assumptions of the government's engineering scheme. Brower's presentation, therefore, focused on the inadequacy of the bureau's geological studies of the dam site, identified miscalculations made by its economists, and demonstrated that its projections of the lake's evaporation losses were suspect. This tactic did more than tip the scales in the Echo Park struggle. It taught conservationists that the most effective way to participate in environmental decision-making was to marshal outside experts and relentlessly scrutinize the methods and presumptions of federal planners.

The victory that saved Echo Park was a crash course in practical politics for Brower and the leaders of the national conservation organizations. It taught them that if citizen lobbies concentrated their activities on a region, or a single resource, their national clout would be limited. In the 1930's "Ding" Darling described conservationists as an "unorganized army." The Echo Park struggle forced them into a phalanx that commanded respect along Pennsylvania Avenue, and it also constituted a political object lesson that led to future coalitions for conservation.

From the Echo Park encounter, Dave Brower learned a lot about himself and about the Sierra Club's potential as a national voice for conservation. His experiences in Washington indicated he had a flair for lobbying, and he realized there was a leadership vacuum that the Sierra Club could fill if it wanted to extend itself. The club did not change its name, but in the aftermath of Echo Park, it became a truly new and different organization. Its membership rolls swelled, new chapters were chartered in fifteen states, and Justice William O. Douglas, in a flush of enthusiasm, briefly joined the governing board of the organization.

In January 1960, David Brower was smarting from a rebuke by the old-guard members of his board, who were shocked by the blunt accusation of malfeasance he had fired at officials of the U. S. Forest Service and the California Highway Commission. In an act of defiance, Brower printed a declaration in the club's bulletin that called for militant action in the coming decade. That article was infused with the passion and the vision that became Dave Brower's trademark. He relished being charged with extremism, confessed that he had an abiding distrust of authority, promised further attacks on misguided government experts, and warned against soft compromises that betrayed cardinal principles. Brower's manifesto was, at once, an open letter to his critics on the board and a call to action for those who shared his outlook about conservation's future. "What [land] we save in the next few years," he intoned with Old Testament fervor, "is all that will ever be saved."

As a gambit to fulfill this new-leadership pledge, Brower urged the club to lobby for a plan to have the federal government buy the Point Reyes Peninsula and designate it a national seashore. He also used his talents to help Dr. Edgar Wayburn, one of the club's most forceful directors, fashion strategies that enabled the Sierra Club to win an eight-year fight to create a Redwood National Park in northern California.

It was the nation's response to the Sierra Club's campaigns and to the subsequent battle it mounted against proposed hydroelectric dams in the Grand Canyon that made Brower one of the giants of conservation history. Events in the 1960's vindicated time after time his perception that it would take an aggressive point man and a militant organization to galvanize the conservation cause. Dave Brower was an environmental activist before the term was coined, and it was his overall strategy that catapulted his club into the vanguard of a new wave of conservation action in this country. Brower loomed above the other leaders because he did so many things so well. A jack of all trades, he excelled as a legislative lobbyist, had no peer as a publicist, and had a charisma that made him one of the best environmental stump speakers of his time.

One example of Brower's versatility was the stunning exhibit-format books he designed and edited for the Sierra Club. Dave persuaded some of the nation's finest photographers and nature writers to collaborate on these tomes. Some volumes were prepared to propagandize for, or against, issues being debated across the nation; and each constituted a powerful statement on behalf of conservation principles. In one hectic twenty-six-month period that commenced in the summer of 1962, David Brower produced four battle books that had an influence on the outcome of great controversies. The forceful messages in *In Wildness Is the Preservation of the World, The Last Redwoods, The Place No One Knew* and *Time and the River Flowing* were carried to the desks of men and women who shaped public opinion and were hand-delivered to members of Congress—at times by Brower himself—who would cast the crucial votes on vital issues.

Later, when the redwood and the Grand Canyon controversies were near a climax, Brower borrowed a technique from the advertising industry and purchased full-page advertisements in some of the nation's leading newspapers. Those first-of-their-kind ads exemplified the extremist approach advocated by the executive director. They demanded better resource stewardship and rudely questioned decisions of presidents and cabinet officers.

An open letter to President Johnson contained this salvo: "Mr. President: There is one great forest of redwoods left on earth; but the one you are trying to save isn't it! . . . meanwhile they are cutting down both of them!" A second emotional ad began with the accusation, "This time they want to flood the Grand Canyon!" and concluded with the question, "Should we also flood the Sistine Chapel so tourists can get nearer the ceiling?"

Dave's innovative newspaper campaigns evoked dramatic responses. They nettled officials in executive suites, they filled congressional mailbags with irate letters, and they persuaded hundreds of citizens to join the Sierra Club by signing coupons attached to the ads. It was Brower's conviction that unless one organization fought unflinchingly for a whole loaf, no half-loaf victories would ever be achieved.

These ads aroused a furor on another level when the White House ordered the Internal Revenue Service to revoke the Sierra Club's tax exemption as a nonprofit organization. That punitive stroke generated a new ripple of sympathy and support for the embattled Sierra Club. *Life* magazine put Brower on its cover and dubbed him "his country's Number 1 working conservationist"; and the *New York Times* hailed the Sierra Club as "the gangbusters of conservation."

David Brower was much more than a daring field general. No one did a better job of spreading the new ecological gospel in the hinterland. Although not a philosopher like John Muir, Dave was an original thinker who developed penetrating insights from ideas that were blowing in the ecological winds of the 1960's. He pioneered the right-to-know, he taught the rudiments of what became known as ecotactics, and he helped start the train of thought exemplified by the bumper-sticker slogan, "Question Authority."

The end of David Brower's Sierra Club career was ironic, and perhaps inevitable. It was ironic because it came so soon after the club won its fights for a Redwood National Park and against the dams in the Grand Canyon. And it was probably inevitable because it involved a leader who was an Isaiah, not a Moses.

Some have sought to attribute Brower's downfall to his excess of zeal, to the bankruptcy-threatening deficits he contracted, or to the mutinous, headstrong stances that caused him to defy his board of directors. All of these were surely factors, but in the end he was brought down by the fact that he was a free spirit who was—and always had been—the antithesis of the organization man. In the end, he was ousted by the votes of admiring friends who wanted a director who would take direction.

David Brower, in his own inimitable way, nursed his grievances for a few hours and then did a remarkable thing for a deposed leader of a national organization. He summoned a few staunch friends and founded a new organization, Friends of the Earth. Brower gave FOE a global focus, and promptly initiated new projects such as a political-action arm for the movement that was named The League of Conservation Voters.

Friends of the Earth quickly issued its own battle books, prepared a provocative activist handbook that merited several printings, and developed a distinctive voice and presence on the national scene. Brower knew how to reach out to the American people. A phenomenal solicitation letter rocketed the new organization's membership from 1,000 to 16,000 in a few months. And a full-page, anti-Super Sonic Transport ad in the nation's major newspapers—"Breaks windows, cracks walls, stampedes cattle, and will hasten the end of the American wilderness"—positioned FOE as the spearhead of the fight that killed this glamorous Big Technology project that had been pushed by two presidents.

It is inaccurate to picture David Brower as a one-man show. He had many collaborators who played stellar roles in his campaigns. But if one believes, as I do, that the American people altered their thinking about their environment in the 1960's, in due course there will be laurels aplenty for David Brower, for his lieutenants, and for the other leaders of the Sierra Club as well.

CHAPTER XVII

Howard Zahniser and the Fight for the Wilderness

Mechanized recreation already has seized nine-tenths of the woods and mountains; a decent respect for minorities should dedicate the other tenth to wilderness.

—Aldo Leopold

We simply need that wild country available to us, even if we never do more than drive to its edge and look in. For it can be a means of measuring ourselves and our sanity as creatures, a part of the geography of hope.

—Wallace Stegner
The Sound of Mountain Water (1961)

W HEN BOTH houses of Congress cast overwhelming votes for a wilderness bill in 1964 it was a triumph for the old generation of conservationists, for a concept nurtured by Thoreau and John Muir. This landmark law also represented a solid victory for the nascent environmental movement and its efforts to place restraints on technology and to inculcate respect for the life-sustaining processes of nature.

The campaign to persuade the American people that it was in the national interest to preserve large enclaves of wildlands was initiated thirty years earlier by an elite group of conservationists who formed the Wilderness Society. The society had its beginning in the fall of 1934 when four friends sat on a riverbank in the Great Smoky Mountains and conceived of an organization dedicated to saving remnants of the nation's virgin lands. They were Benton MacKaye, Harvey Broome, Bernard Frank, and Robert Marshall.

Marshall, the thirty-three-year-old human dynamo who personified the Wilderness Society in its early years, was once described by Justice Benjamin Cardozo as a "radiant presence." His short life was filled with physical and intellectual adventures. Bob was a latter-day explorer whose prodigious walks into wild country ranged from Alaska's Brooks Range on the Arctic Circle to the upper Sonoran Desert at the edge of Arizona's White Mountains. When he died at thirty-nine, Bob Marshall had lived with Eskimos and written a best-selling book about their culture, had composed a volume advocating public ownership of all U.S. forestlands, and had become an advocate of civil liberties and Norman Thomas's brand of socialism.

But his great passion was conservation, and his love for wild country was revealed by his response to a skeptic who asked how many acres of wilderness the nation needed. Marshall replied: "How many Brahms symphonies do we need?"

Born into a prominent, wealthy family, Bob developed, as a youngster, a fascination with wild things on hikes into vastnesses of New York's Adirondack Forest Preserve. His restless interest in the natural world subsequently ripened into a career in forestry, and by the time he was thirty he was a national expert in the recreational use of forestlands. In 1932, Gifford Pinchot asked Marshall to help him outline a new federal forestry policy for President-elect Franklin Roosevelt, and a short time later Secretary Harold Ickes assigned him the job of managing the Interior Department's Indian forests. As an executive, Bob Marshall spent most of his time in the field, and his backcountry walks in the West showed him that machines were poised at the edge of the nation's last wildlands.

The idea that federal foresters should take action to protect enclaves of unspoiled country in the nation's publicly owned forests was formulated two decades earlier by Arthur Carhart, the first landscape architect hired by the U.S. Forest Service. He argued that some "unmarred" lands within the National Forests should be placed in reserves as "the property of all people." Art Carhart's concept became a reality a few years later when Aldo Leopold persuaded his superiors to designate an expanse of the Mogollon Mountains in New Mexico as the nation's first semiprotected "wild" area.

But by 1934 experienced land managers like Bob Marshall and his friends knew that such regulations were, at best, tentative decisions that could be readily cancelled or modified by subsequent administrators. Moreover, they were alarmed that growing demands for scenic parkways and relentless pressures by developers who wanted access to minerals and timber would slowly shrink the nation's estate of wildlands unless the boundaries of all reserves were delineated by laws. Like Muir before him, Marshall was convinced that "legislative interference" was imperative if "the tyrannical ambitions of civilization to conquer every niche of the whole earth" were to be repelled.

In the cramped days of the Great Depression, Marshall, Leopold, and the other founders of the Wilderness Society must have sensed that they had embarked on what might be a quixotic quest. In the first years of its existence, the society sought to publicize the values of wilderness, and it functioned to a large degree as an extension of Bob Marshall's pocketbook and personality. And it was a generous bequest in his will that enabled the society to survive during the war years of the 1940's.

In 1946, six years after Marshall's untimely death, Howard Zahniser picked up the torch of Bob's crusade when he left his position in the federal government and became the paid executive of the Wilderness Society's one-man operation in Washington. Some surely regarded the bookish, low-key Zahniser as a poor choice to spearhead a national campaign for a radical reform. Zahnie was a reserved man, but he had attributes that made him an ideal person to lead the dogged, eighteen-year effort that ensued. He had an uncommon capacity for friendship, he wrote poetry, he adored Thoreau, and he had an affinity for the Book of Job. These were traits and interests that gave Howard Zahniser the faith and patience he needed to continue his seemingly unending walks in Washington's corridors of power.

In his first years at the helm of the society, Zahnie acquired vital insights into wilderness values on hikes in the same Adirondack expanses that had earlier fired the imagination of Bob Marshall. Moreover, when he presented testimony to the New York legislature in support of the "forever wild" covenant in that state's constitution, Zahniser formulated in his mind some of the basic concepts he later incorporated into the initial wilderness bill he submitted to his friends in Congress.

There was also a fascinating overlap in the lives of Rachel Carson and Howard Clinton Zahniser. Born fifteen months apart, they grew up in small towns in the valley of Pennsylvania's Allegheny River. Each graduated from a small college. Each taught school for a season before being employed to publicize the programs of the U.S. Fish and Wildlife Service. And each ultimately left government service (Rachel after sixteen years, Zahnie after fourteen): one to write books, the other to devote the remainder of his life to a crusade for the wilderness cause. And through a final coincidence, the lives of these environmental siblings ended a few days apart in the spring of 1964. Cancer claimed Rachel Carson on April 15; Howard Zahniser succumbed to heart failure twenty-one days later.

The Wilderness Society's 1950 decision to play a dominant role in the fight against the Echo Park Dam put Howard Zahniser at the nerve center of what became the most important exercise in conservation politics in over two decades. It also furnished a forum where he could collaborate with other conservation leaders, study the behavior of individual congressmen, and learn how to muster support for a controversial conservation proposal.

Although he admired Dave Brower's frontal attack on the Bureau of Reclamation's proposed dams, Zahniser readily recognized that it would take a different strategy to persuade Congress to pass a law that cut against the grain of the prevailing American concept of resource development. His adversaries, as he well knew, viewed his wilderness legislation as an idea advanced by elitist recreationists who wanted to "lock up" resources that would be needed by future generations.

When Zahniser drafted a wilderness bill in 1956, the 7,000-member organization he led was embarking on a long-shot political undertaking. The preservation ethic he espoused ran counter to the mainstream values of the 1950's. With the Forest Service and the National Park Service taking the stance that wilderness legislation was unnecessary, the daunting question Zahnie faced was whether the nation's smallest conservation organization could persuade the Congress of the world's most development-oriented country to pass a law declaring that the resources in tens of millions of acres of its national lands were never to be developed.

Zahniser sensed early on that a congressional struggle over wilderness legislation would reverse the normal posture of the contending parties, with his society defending its proposal against frontal attacks launched by antagonistic industry groups. This convinced him that to overcome the powerful economic arguments of the mineral, timber, and grazing interests who were arrayed against him he had to fashion pragmatic arguments that would bring outdoorsmen, editorial writers, some prominent Western senators—and perhaps even a President—to his side.

Howard Zahniser presented testimony at nine sets of public hearings over an eight-year period, and this gave him ample opportunity to develop new themes and fresh arguments to rebut the tiresome contentions of his opponents. Zahnie evolved a cogent set of arguments about the multiple uses and services that the wilderness offered to Americans. To disarm his critics, his presentations were low-key. He pointed out that a reserve of untrammeled lands would serve as the most important watersheds for river basins, as the locale of the finest high country fishing, as incomparable refuges for wildlife, and as priceless gene-pool laboratories where future scientists could go to find solutions to medical and horticultural problems. He also devised answers to the lock-up argument that appealed to so many mining-state congressmen. This took the form of an amendment which gave presidents a statutory "key" to unlock any reserve if a specific resource was needed in a national emergency.

Howard Zahniser was the right lobbyist for the Wilderness Society because he exuded goodwill and had the perseverance to outwait and outwit his adversaries. Although he saw his wilderness bill watered down and rewritten sixty-six times, his faith never faltered. There was mutual admiration between the militant Brower and the patient, soft-spoken Zahniser. The temperaments of these two men differed, but they shared a love affair with the earth's sacred places and each gained insights through their common campaigns.

The Wilderness Society faced heavy odds when it launched its campaign for wilderness legislation in the winter of 1957. No other nation had taken such a radical step, and Colorado's prickly Wayne Aspinall (who would soon hold this legislation hostage as the autocratic chairman of the House Interior Committee) probably expressed the prevailing sentiment in the West when he characterized the original bill as "a crazy idea."

The front-line opponents of the bill were the mining and timber and grazing interests who, since Pinchot's time, had been the dominant and domineering users of National Forest lands. The Eisenhower administration was attuned to those user groups, and the Forest Service and the Park Service, agencies that should have been all-out supporters of the wilderness ethic, took the position that such legislation was superfluous.

Howard Zahniser knew from the start that his cause would not prevail unless he was able to persuade a powerful Western senator to be its champion. But it was important to begin, and Zahnie began by recruiting a Midwesterner, Senator Hubert Humphrey of Minnesota, and a Teddy Roosevelt Republican from Pennsylvania, Representative John Saylor, to serve as co-sponsors of the first bill.

Zahniser's record as a team player put him in a strong position to revive the Echo Park coalition and to encourage the other conservation organizations to make the wilderness fight their own. The many friendships he had nurtured strengthened his ability to lead this seemingly lost-cause campaign. And Zahnie had stalwart allies in Justice William O. Douglas, Joseph Wood Krutch, and Sigurd Olson, whose pro-wilderness books and essays served as background music for a steady flow of articles Howard composed to bring the wilderness gospel into the mainstream of American thought.

The first showdown came in the Senate Interior Committee during the 86th Congress in 1959. The result was a legislative dogfight that dramatized the profound disagreements among senators representing the Western states. Obdurate senators from Colorado and Wyoming first offered a meaningless substitute and followed with a dilatory "committee filibuster" of endless amendments. This stalemate sent a signal to the society that its four years of lobbying had been wasted.

Irrepressible as always, Zahnie viewed this fiasco as a learning experience, not a defeat. He had watched the machinations of his opponents with a shrewd eye, and this period's final skirmishes told him that New Mexico's Senator Clinton P. Anderson, the prospective new chairman of the Senate Interior Committee, was the leader he was looking for. Clint Anderson was one of the Senate's best strategists in a floor fight, and, more important, he had acquired strong convictions about wilderness values thirty years earlier from an Albuquerque friend named Aldo Leopold.

The first turning point in the fight came after the 1960 election when Anderson agreed to lead the wilderness crusade and instructed his staff to prepare an "Anderson Bill" which he would introduce when the new Congress convened. The second occurred on February 13 when Chairman Anderson persuaded President Kennedy to feature an endorsement of his bill in the special conservation message he was planning to transmit to Congress. These maneuvers produced a drastic change for conservation. Now, for the first time, the wilderness alliance had the support of a President and a floor leader who knew how to muster majorities on Capitol Hill.

Once he had committed himself, it was characteristic of Clinton Anderson to pull out all the stops. Early hearings were scheduled, and by the end of July 1961 he had marshalled the votes needed to report his bill. The conservation coalition responded, and by midsummer mail to members of Congress was running sixty-six to one in favor of the Anderson bill.

Detained in Albuquerque by surgery, Anderson persuaded Frank Church of Idaho, a first-term senator from a state whose user industries were violently opposed to any "lockup" of undeveloped lands, to serve as the floor manager of his legislation. Church relished this adventure in high-risk politics, fought off crippling amendments, and listened, amazed, as his colleagues voted for the Wilderness Act by a seventy-eight to eight majority.

But euphoria was premature. Though samplings of sentiment indicated that if the members of the House of Representatives were allowed to vote, they would vote overwhelmingly for the Senate bill, the Chairman of the House Interior Committee, Wayne Aspinall, personally prepared a blockade to prevent a vote on the Anderson bill. It would take three years of wrangling—and John Kennedy and Howard Zahniser would both be dead—before the wily Colorado congressman would relent and permit the House to work its will on this pioneering legislation.

The autocratic Aspinall, who had all of the good and bad traits of an industrious hedgehog, was Howard Zahniser's nemesis for five full years, from 1959 to 1964. A masterful manipulator, he was one of the last of this century's chairmen to run his committee as though his vote was the only one that counted. Moreover, he was contemptuous of the lawmaking skills of most senators; he resented intrusions on "his prerogatives" by the executive branch. He kept the members of his committee in line by constantly reminding them that he was in control and would personally decide the fate of every bill that was referred to his committee.

Aspinall's hedgehog qualities came into play any time "unreasonable outsiders" criticized his slow pace or intimated he was obstructing action on important legislation. He would roll himself into a self-righteous ball, flourish his spines, and complain that lazy senators saddled him with their detail work. Then, as he retreated into his burrow, he would make a mental note to shelve the legislation favored by members of Congress who had the temerity to criticize him.

Although Wayne Aspinall tried Zahnie's patience in a hundred different ways—and spent five years thwarting his hopes— like Job, Zahniser never responded with rancor. This demonstration of brotherly love evoked such amazement that when Zahniser's death helped bring his remarkable tableau to its denouement, some of us wondered whether a catharsis of shame prompted Aspinall to relent and allow the Wilderness Bill to come to a vote in the House. In any event, the legislative stalemate was finally broken in the summer of 1964 when, in a moment of mutual exhaustion, Anderson and Aspinall struck a deal that allowed a watered-down wilderness law to be sent to the President.

Those who remember Zahnie's ministry in Washington will never forget his playful mind nor his rich and generous spirit. When the long-precarious condition of his health was revealed at his death, one sensed that the love of life he wore on his sleeve was embedded in an awareness that he was living each day on borrowed time.

No one viewed it in that light at the time, but the 1964 Anderson-Aspinall compromise was, in truth, a political wager. The concession the Colorado chairman demanded and got was an amendment that no lands could be added to the newly created Wilderness System unless both branches of Congress voted for such additions. Representative Aspinall was betting that this provision—and the twenty-year procedural steeplechase it erected—would sap the strength of the wilderness movement and give the congressional delegation from each Western state a veto over any future expansion of wild areas in their state.

A weary Senator Anderson had a contrary hunch that if his bill became law, wilderness activism was here to stay and would flourish even in the West. Nearly a quarter of a century later, history tells us that Clinton Anderson not only won his wager, but that the idea he and his colleagues championed struck a note of restraint that gave credence to one of the newborn causes of ecology.

CHAPTER XVIII

Science, Law, and Environmental Reform:

THE ENVIRONMENTAL DEFENSE FUND BLAZES A TRAIL

In no other political and social movement has litigation played such an important and dominant role. Not even close.

—DAVID SIVE
Pioneer Environmental Lawyer

At first we were taken for a bunch of bird-watchers, but the industry is worried now. They've got the money, we've got the science.

—DR. CHARLES F. WURSTER
EDF Scientist (1968)

ALTHOUGH *Silent Spring* started a fundamental shift in thinking about technology and the environment, it did not offer a battle plan to roll back the dissemination of DDT. Even as Rachel Carson was finishing her book, sales of the new toxicants were accelerating, and research scientists working for U.S. chemical companies were concocting poisons more potent than DDT.

The first effective campaign against DDT commenced in 1966, two years after Miss Carson's death, when a lawsuit was filed in a county court to stop the spraying of DDT to control mosquitos on Long Island. The Suffolk County Mosquito Control Commission (SCMCC) was the defendant in that class action. The symbolic class plaintiff was a Long Island housewife, Carol Yannacone. Her lawyer husband, Victor, prepared pleadings alleging that SCMCC's DDT was killing fish and wildlife and injecting long-lived poisons into foods eaten by Suffolk citizens.

Victor Yannacone, then in his twenties, was a combative Patchogue attorney who quickly realized that his courtroom crusade would flounder unless he got help from scientists who knew hard facts about DDT. His search for scientific talent became an exercise in serendipity when he established contact with a local association of conservationists, the Brookhaven Town Natural Resources Committee (BTNRC).

The citizens and scientists of BTNRC were alarmed by the damage DDT was inflicting on their environment. Among the volunteers who helped Yannacone press his counterattack against SCMCC's spraying were Charles F. Wurster, assistant professor of biology at the State University of New York at Stony Brook; George M. Woodwell, a scientist at the Brookhaven National Laboratory who would later part company with the Atomic Energy Commission and become one of the preeminent ecologists of this century; Dennis Puleston, Brookhaven's Director of Technical Information; and Arthur P. Cooley, a biology teacher at Bellport High School.

Wurster had just completed a two-year program of research proving that large numbers of birds were exterminated when DDT was sprayed on elm trees, and he was outraged by assertions of the SCMCC Commissioner that DDT was harmless. Woodwell had published a paper demonstrating that DDT persisted in forest soils. And Puleston, an ornithologist, was involved in an experiment to ascertain whether DDT was implicated in the decimation of the ospreys on nearby Gardiners Island.

With Yannacone as their ringmaster, BTNRC's pickup team of moonlighting scientists (once described as "Rachel Carson's army") soon became tigers in the courtroom. More important, the evidence this team presented to the judge—and the publicity it generated—got results. A temporary injunction stopped the Suffolk mosquito men from spraying DDT, and political ripples from the trial swayed Suffolk's elected officials to ban the use of DDT in their county.

Yannacone was a brilliant tactician, but from the beginning he had no illusions that litigation would produce resounding legal victories. His maverick motto was "Sue the Bastards," and he envisioned his lawsuits as show trials to dramatize environmental truths that would ultimately compel members of the legislative and executive branches of government to act. He was willing to lose court decisions if his cause prevailed in the court of public opinion.

When the initial courtroom adventure was completed, Victor Yannacone and his colleagues were excited by the prospect that their collaboration had scored points for environmental reform. But none of them dreamed in 1966 that their mosquito case would start a rollback that would culminate six years later in a national ban on the use of DDT.

Emboldened by the results of the SCMCC case, on October 6, 1967, the Yannacones and eight scientists gathered to sign a certificate incorporating the Environmental Defense Fund (EDF). Their charter spoke of lofty national aims, but the basic goal was to thrust environmental issues into the courts. The EDF began barefoot with no officers, no staff, no money, no office, and no members. But it possessed a cadre of original minds and a conviction that scientists and lawyers could use the legal system to aid the cause of environmental protection.

The minutes of the initial meeting of EDF's trustees reveal that its incorporators were in a semischizoid frame of mind. First, there was a unanimous vote that the organization should be mindful of its precarious position and "proceed with caution." A few minutes later another unanimous resolution authorized the prompt filing of a double-barreled lawsuit against the Michigan Department of Agriculture (MDA) to halt the spraying of dieldrin on a supposed "infestation" of Japanese beetles and against nine Michigan municipalities to prevent them from spraying DDT on elm trees.

Within a week, Yannacone and Wurster were in Grand Rapids with a complaint supported by voluminous affidavits seeking a federal judge's signature on a temporary injunction. As expected, EDF lost its first courtroom round against the MDA's sprayers. But once again the publicity generated by the scientific evidence of EDF's experts alerted Michiganders to the threat long-lived pesticides posed to their fish and wildlife and especially to the popular new coho salmon fishery in Lake Michigan. The Michigan foray came to a sudden climax when the Federal Food and Drug Administration validated EDF's case by seizing a shipment of coho salmon that was contaminated by residues of DDT and dieldrin. The furor that followed not only forced the MDA to backtrack, but made Michigan the first state to take strong action against the use of DDT and its deadly cousins.

EDF's next challenge involved filing a joint petition with local conservation organizations requesting that the state of Wisconsin declare DDT a pollutant that violated its water pollution law. That petition produced the most exhaustive hearing ever held on the pesticide issue in the United States. Twenty-seven days of testimony made this proceeding a *cause célèbre* that captured the attention of scientists all over the world.

The Wisconsin case shifted the pesticide argument away from generalities to a sharp, specific question: was DDT a pollutant under Wisconsin's water quality laws? The hearing room in Madison became the scene of a confrontation where each side could present its best evidence and cross-examine the opposition's experts. Here was a neutral forum where, with the nation listening, scientists had to play hardball and meet their antagonists with facts supported by actual research.

Yannacone and Wurster sought to convert that hearing into a wide-ranging science seminar that would summarize the case against DDT. The indefatigable Wurster drew his witnesses from a pool of more than one hundred prominent experts he had enlisted to serve on EDF's Scientists Advisory Committee. Testimony was presented in the fields of ecology, botany, limnology, chemistry, pharmacology, entomology, and wildlife biology by eminent scientists who related their recent findings about DDT to Wisconsin's water and the web of life of its ecosystems.

Confronted with the holistic framework adopted by those experts, the strategists of the agri-chemical industry elected to take a stance that belittled ecology. The EDF witnesses, the industry's lawyers argued, were specialists who were "out of their field" in attempting to draw conclusions about environmental impacts. EDF's riposte made ecology the pivotal issue of the proceeding. How, Victor Yannacone queried, could society deal with the insidious encroachments of DDT unless it heeded scientists qualified to generalize about the effects of such agents on the overall environment? Wurster and Yannacone knew their strategy was working when the Wisconsin hearing examiner told a reporter, "A legal case is like a wall, and you have to put it in brick by brick. Usually the scientist has been interested only in his own brick. Now ecologists are trying to put all of the bricks together."

As a scientific showdown, the decision of the Wisconsin pollution agency against DDT was a milestone for the Environmental Defense Fund. It verified the assumption of EDF's founders that a partnership between law and science could thrive in a courtroom environment. It demonstrated that there was an environmental science community in the United States ready to fight for ecological reforms. And it showed that if individuals were willing to make sacrifices, it was possible for a small organization to hold its own against the big battalions of the chemical industry.

Sacrifice by individuals made the great days of EDF's first phase possible. Although Michigan and Wisconsin conservation groups advanced funds to pay court costs and travel expenses, it was the willingness of Yannacone, Wurster, and others to work without compensation in far-off places that enabled the Environmental Defense Fund to blaze a trail for environmental law in the 1960's.

Charlie Wurster, a freshman professor at twenty-eight, exemplified the dedication and *esprit* of EDF's corps of scientists during the lean years. He orchestrated the testimony at the Wisconsin marathon, and his incessant correspondence created a network of experts that gave EDF scientific credentials that no other organization had in the 1960's. Wurster updated Rachel Carson's research by acquiring and analyzing the latest pesticide studies from around the globe. During the Wisconsin hearing, George Woodwell paid his colleague this compliment: "Charlie is very, very bright and remembers everything he reads. I think, unquestionably, he knows more about persistent pesticides than anybody in the world."

In the final weeks of 1969 EDF took steps that paved the way for a history-making breakthrough in environmental law. It presented petitions to the secretaries of Agriculture and HEW alleging that DDT was a cancer-causing substance and requested the establishment of a zero tolerance for DDT in human foods. Those petitions resulted in a flurry of environmental lawsuits and a marathon five-month administrative hearing which ultimately established a precedent that citizens could use in the courts to enforce federal health laws. That history-making legal strategy came to a culmination in June 1972 when William D. Ruckelshaus, Administrator of the new Environmental Protection Agency, issued an order terminating the use of DDT in the United States, a triumph foreshadowed by a 1970 New Year's Day decision by Wisconsin officials banning all uses of DDT in that state.

The swift ascendancy to national prominence of this shoestring organization of unpaid volunteers stands out as one of the most remarkable events of the environmental revolution. Moreover, the publicity created by EDF's drive against DDT helped generate political pressure that prompted President Richard Nixon to sign into law, also on the first day of 1970, a far-reaching National Environmental Policy Act (NEPA).

In only two years the Environmental Defense Fund and its unfunded amateurs had come of age. At the end of 1969, a writer for *Science* magazine, the official organ of the American Association for the Advancement of Science, observed with astonishment that ecology was evolving into a "rapidly developing glamour science." But it was not glamour that dominated the outlook of EDF's leaders as they contemplated the 1970's; it was what they had learned about the catalytic influence of science and law on policy-making. They had learned, for example, that if teams of scientists and lawyers did their homework they could shake up bureaucracies and force high-level decision makers to act. And they had taught the American people that informed ecologists were guardians of the public's health and environment.

The waves of national publicity generated by the campaign to ban DDT enabled EDF in 1970 to become financially independent by enlisting 11,000 dues-paying members through direct-mail solicitations. Except for the similar takeoff of Brower's Friends of the Earth, the conservation movement had never witnessed such an explosion of support for a new organization. By 1972, EDF had 36,000 contributing members, an annual budget of $678,000, and branch offices in Washington, D.C., and Berkeley.

The avalanche of support for environmental law enabled EDF's trustees to hire an executive director, to recruit a full-time staff of eight lawyers, six scientists and two economists, and to rapidly become a unique public interest organization. By 1973 EDF had eighty active cases in all parts of the country including a struggle to protect a wild river in Arkansas, an effort to stop construction on a barge canal across northern Florida, a NEPA lawsuit to require refiners to produce lead-free gasoline, and another NEPA action to stop an ill-planned scheme to build a pipeline for hot oil across the tundra of Alaska.

As a result of the Environmental Defense Fund's pioneering in the 1960's, science-law coalitions became a potent catalyst of environmental reform. EDF, with its extended family of science advisors and several hundred volunteer attorneys, inspired the formation of public law groups in all parts of the country. Moreover, its successes encouraged the Ford Foundation to support the formation of a twin organization, the Natural Resources Defense Council (NRDC), and inspired the Sierra Club to enlarge its Legal Defense Fund.

Today, environmental law is part of the fabric of American life. It impacts on policy-making at all levels of government, and EDF and NRDC not only have a powerful voice when new laws are written, but they monitor and guide the enforcement and implementation of the nation's environmental laws. Interdisciplinary litigation is still the forte of these organizations. But environmental law is still evolving, and recent signs suggest that in the future both entities intend to use the talent they have assembled to help the nation—and the world—develop innovative solutions to today's toughest environmental problems.

The NRDC's John Adams and Thomas Cochrane startled arms-control experts in Washington and Moscow in 1986 when they launched a campaign to stress new technologies that now make it feasible for these superpowers to enter into verifiable agreements that would halt underground nuclear tests. As this book goes to press, EDF's Bruce Rich is directing a global effort to save the world's remaining rain forests by persuading the World Bank and other international lending agencies to stop funding projects that are destroying the ecosystems of nations and regions.

In 1988, the work of these teams of scientists, lawyers, and economists represents the cutting edge of efforts to achieve patterns of cooperation on nature's behalf among those who share this planet and its resources.

CHAPTER XIX

Widening the Circle of Ecological Awareness:

RALPH NADER, BARRY COMMONER, PAUL EHRLICH, AND EARTH DAY

To bring environmental logic into contact with the real world we need to relate it to the overall social, political, and economic forces that govern both our daily lives and the course of history.
—BARRY COMMONER, 1971

Human values and institutions have set mankind on a collision course with the laws of nature.
—PAUL EHRLICH, 1972

ANOTHER development that widened ecological awareness in this country was the appearance of a few exceptional teachers who dramatized and popularized environmental issues for the American people. Three individuals who became instructors-to-the-nation on television and through their writings and lectures were Ralph Nader, Barry Commoner, and Paul Ehrlich. Nader was a self-styled consumer advocate whose convictions about corporate responsibility and consumer rights made him a powerful champion of environmental protection. Commoner and Ehrlich were scientists who had an uncommon ability to communicate with a new generation of Americans.

Ralph Nader

The emergence of thirty-two-year-old Ralph Nader as the nation's chief champion of consumers' rights is an improbable episode of American public life. When he vaulted into the national news at a 1966 Senate hearing, Nader was a Washington outsider whose only real contribution to the common weal had been documenting Detroit's disinterest in automobile safety in a little noticed book entitled *Unsafe at Any Speed*.

General Motors, the largest U.S. corporation, had provoked the Senate's inquiry by hiring a private detective to harass and discredit Nader. Had GM's executives been less paranoid—and more receptive to constructive criticism—they might have had the good sense to forego hiring a sleuth and instead telephoned their Chevrolet dealer in the small town of Winsted, Connecticut, to find out what he knew about Ralph Nader and his family.

Such a call would have told them that Ralph's father, Nathra, a Lebanese immigrant, was a thrifty, upstanding member of the community who had owned and operated a successful bakery-restaurant business for thirty years. They might have learned that his friends and customers knew Nathra Nader as a person who was willing to do battle for his belief in the American dream and who instilled in his children the idea that they had an obligation to work to make the system more just and equitable. They might also have learned that Ralph, the youngest child, was a very bright, somewhat offbeat individual who had worked in his father's store during his early years, had graduated from Princeton and then Harvard Law School, had failed in an effort to establish a law practice in Hartford, and was at the time working on some unnamed project in Washington.

But such a call was not made. And unwittingly, GM created a televised morality play that gave this unknown kid from Winsted an opportunity to play an innocent David opposite their corporate Goliath. There was something comically ironic about the whole affair.

Henry Ford came to hate Ralph Nader, and GM's boss, James Roche, once scathed him as "one of the bitter gypsies of dissent." Yet Nader was nurtured on the concept of free enterprise. All he really wanted was for Detroit's auto men to build safer, cleaner, more efficient cars, and to help the U.S. auto industry please its customers and maintain its hegemony in the world marketplace.

The 1966 Senate hearing and GM's out-of-court settlement of Nader's invasion-of-privacy lawsuit would normally have been the end of this drama. However, once he was lifted onto the national stage, Nader used his repertoire of complaints to improvise so many provocative Ralph-the-reformer-comes-to-Washington scenarios that before the curtain was drawn on GM's playlet, the nation had a new Washington-based show called "Citizen Nader."

Ralph's show was sustained by his ever-expanding agenda of reform which reached far beyond the problems of the automobile industry. He was convinced that American industry had lost its way, had become so smug, so engrossed in a single-minded quest for profits, that it was needlessly eroding the quality of life in this country. Ralph Nader, who rarely entertained modest aims, envisioned himself in 1967 as an evangelist with a mission to censure industry's backsliders and to lead a national crusade to achieve a "qualitative reform" of the industrial revolution.

General Motors made Nader a national figure, but his success as an ombudsman for American consumers was due to his exemplary integrity, and to a panache that made him the designated stone thrower at Washington's Goliaths. His evident willingness to lead battles against still other giants galvanized frustrated folk in government and industry who shared his convictions, and their calls and revealing documents quickly turned Nader's Washington apartment into the nerve center of a campaign to make industries accountable for their sins and shortcomings.

With the assistance of a small staff of ill-paid idealists, Nader was soon orchestrating skirmishes on many fronts. Some of his initial interventions, for example, involved demands for the recall of unsafe autos, inquiries into the operation of our network of natural gas pipelines, complaints about impurities in foodstuffs, presentations about the pollution of the nation's water supplies by phosphate detergents, demands that action be taken to improve safety practices in coal mines and provide federal benefits for miners who were victims of a black-lung epidemic, and the unveiling of evidence that the Federal Trade Commission was indifferent to industrial sloth.

Nader had no training in public relations, but he had political instincts that made him one of the master publicists of his generation. He and the television revolution arrived on the national scene about the same time, and he quickly surveyed the scene and rode the crests of the new electronic waves with great dexterity. No one in Washington was better at packaging facts or simplifying complex issues than Nader and the young researchers he trained (soon known as Nader's Raiders). A typical Nader press release of the 1960's both made the evening news and put business executives on the defensive.

Ultimately, what gave Nader a unique credibility was the transparent purity of his motives. His selflessness blunted the arrows of his critics and bound his followers to him with hoops of steel. Ralph Nader was a knight who needed no armor. His constancy and the depth of his commitment were certified daily by his austere lifestyle and by the circumstance that all of his lecture fees—and the entire $425,000 settlement he won from the General Motors corporation—went into the revolving fund that supported his young staffers and his projects.

Although environmentalism was already making headway when Ralph Nader appeared on the national scene, his efforts on behalf of consumers helped to quicken the nation's understanding of ecological issues and resulted in fruitful interaction between the two new movements. Some of Nader's winnings from his encounter with General Motors were plowed back into the establishment of Public Interest Research Groups at several universities, and into a trenchant series of baseline environmental investigations of toxic substances, air and water pollution, and land use. This and similar efforts widened the nation's understanding of ecological issues and served as an object lesson on how best to mold and mobilize public opinion behind common goals of both groups.

From the start, Nader's program included more than the quality of the products offered by industry to American consumers. Once he looked at people as consumers of the air and water in their everyday environment, it was inevitable that Nader's agenda of corporate reform would encompass the effluents emitted by smokestacks and the tail pipes of automobiles. This concern spurred Ralph to spearhead the fight for legislation with enforcement capabilities that would enable environmental administrators to clean up the nation's air and water.

There were other important Nader initiatives that enlarged the common goals of these nascent movements. To Nader, the issue of working conditions in factories and mines was a paramount environmental problem. Tens of millions of citizens spent a large portion of their lives in unhealthy workplaces, and Ralph reproached the conservationists for wearing blinders that prevented them from seeing this vital environmental issue.

But Nader himself had blinders that limited his interest in the natural world and probably obstructed more fruitful cooperation between two inherently overlapping causes. Ralph kept the conservationists at arm's length because he was never really comfortable with them. The world he inhabited was a glassed-in space where people were obsessed with searching for facts and translating ideas into action. One of his associates once observed that Ralph Nader was more interested in the ingredients of a hot dog than in the contamination of the earth. There was no sign that he had ever read the writings of Thoreau or Muir or Aldo Leopold, and he evinced no interest in watching birds, taking walks in the woods, or peering into tide pools. Strangely, this boy from the Berkshire Hills never developed a love of the out-of-doors.

But despite this blind spot, Ralph Nader must be ranked as one of the most effective nongovernmental leaders of his generation. In 1988, admiring environmentalists concur in the judgment biographer John McCarry pronounced fifteen years ago: "If technology is a church, as many of Nader's enemies and perhaps he himself believe, then Ralph Nader is its first saint."

Barry Commoner

Barry Commoner, the inquisitive son of Russian immigrants, grew up in a poor section of Brooklyn, ultimately earned a doctorate in biology at Harvard, and found himself in the 1950's on the faculty of Washington University in St. Louis. Barry's initial interest in environmental pollution was aroused by his concern over the effects on human health of radioactive fallout from the nation's bomb testing in Nevada. "It was," he observed long afterward, "the AEC [Atomic Energy Commission] that turned me into an ecologist."

St. Louis was downwind from the Nevada Test Site and alarm over the appearance of strontium 90 in local milk caused a group of scientists and physicians to form a Committee for Nuclear Information (CNI) to investigate the inroads of radioactive isotopes into their environment. The tireless Commoner became a driving force within CNI, and by the end of 1958 it had emerged as the nation's first citizen-led environmental health organization. CNI operated from a conviction that citizens had a right to know the facts about fallout, and Dr. Commoner and his colleagues sensed at the outset that the Atomic Energy Commission's experts were cozily concealing ominous information from the American people. CNI responded by launching a series of its own radiation research projects.

One long-range CNI study involved the analysis of 200,000 baby teeth collected from St. Louis children to measure the uptake of strontium 90 fallout into human bone. A second study—an assessment of the environmental consequences of an AEC scheme to use a hydrogen bomb to excavate a new harbor in northern Alaska—demonstrated that radiation residues would poison the subsistence foods of Eskimos.

A third project generated an international shock wave when CNI scientists organized a real-world interpretation of hypothetical information that civil defense "experts" had presented to congressional committee. This effort described the destruction that would occur in the event of a nuclear war and then used the data to project the outcome if a hydrogen bomb were dropped on St. Louis. This study had worldwide reverberations because it was the first systematic effort by scientists to compel people to fathom the consequences of an all-out nuclear war.

Commoner summarized what he had learned during his CNI work in a 1966 book, *Science and Survival*. With one eye on the scientific establishment, he observed that, "The age of innocent faith in science and technology may be over." And with his other wary eye on the burgeoning environmental community, he issued a challenge to create "a new conservation movement . . . to preserve life itself."

The high point of Commoner's contribution as an ecology instructor for the nation came with the publication of *The Closing Circle* in 1971. Like *Silent Spring* this book won a huge audience when it was serialized in *The New Yorker* magazine.

The Closing Circle led millions of readers into the dense thicket of ecology and taught them how to pick their way out by thinking environmentally. Like Carson, Barry knew how to use case histories to dramatize the inexorable functioning of nature's laws. His book clearly explicated the poisoning of the air in Los Angeles, the pollution of farmland in Illinois, and the cycles of contamination that were destroying life in Lake Erie.

Diagnosis was Commoner's forte, and he concluded that the technological mistakes that had produced the environmental crisis were the logical outcome of "the social mismanagement of the world's resources." He believed that society could only escape the predicament it had created by adopting ecological reforms that would close the circle gradually and bring the whole human enterprise "into harmony with the ecosphere."

The Closing Circle was a crisp brief for new policies that would both preserve nature's ecosystems and sustain the quality of life for earth's inhabitants. Its author wisely perceived that it would take time for "collective social action . . . [to] change the course of history." And he took pains, in the final pages, to point out that there was no panacea which would solve the ecological impasse: no individual or committee, he wrote, "can possibly blueprint a specific 'plan' for resolving the environmental crisis."

But Commoner, who had a tendency to think of himself as environmentalism's one true prophet, nursed a growing discontent with the tedious compromises that emerged from the give-and-take of our political system. As a consequence, after the 1973 oil embargo, Barry turned away from science and suddenly became a peddler of political panaceas. *The Poverty of Power,* his 1976 book, contained ecological rhetoric, but it was, in essence, a statement by a self-appointed expert in energy economics. The new Barry Commoner was a social architect who was selling solar energy as a nostrum that would change the course of history and allow the United States to wriggle out of OPEC's chokehold.

Commoner's passage from ecology into politics was abrupt and complete. *The Politics of Energy*, his next book, was a political statement. It served as a springboard for his quixotic personal campaign for the White House in 1980, which, in turn, led to his withdrawal as an active environmental leader.

Paul Ehrlich

Like Barry Commoner, Paul Ehrlich emerged from obscure beginnings. The son of a salesman and a public school Latin teacher, Paul spent his youth in Maplewood, New Jersey and did his graduate study in biology at the University of Kansas. He was a professor at Stanford when he made his meteoric entrance into the environmental firmament in 1968 with a terse book, *The Population Bomb*, published under the auspices of the Sierra Club.

In composing this sensational work, Ehrlich was not breaking new ground. The Reverend Thomas Malthus's theory that unrestrained increases in population represent a threat to earth's inhabitants had been nagging at conservationists since William Vogt made a bold restatement of the Malthusian thesis in a 1948 work, *Road to Survival.*

Dave Brower expressed the consensus of the environmental movement on the subject in 1966 when he said, "We feel that you don't have a conservation policy unless you have a population policy." Brower later persuaded Ehrlich to write *Bomb*, and both were astounded when it spread like autumn leaves across the nation's campuses and, with the sale of three million paperback copies, became the most widely read ecology book of the 1960's.

It was, in part, the overheated atmosphere of the Vietnam era that made it possible for a young scientist whose only previous work bore the title, *How to Know the Butterflies,* to catapult himself onto the national stage. Nineteen sixty-eight was an auspicious time for *The Population Bomb* to appear, for millions of students were convinced that their elders were devaluing their lives and ignoring ominous new truths about the fragility of earth's ecosystems. Ehrlich's extravagant claims made his book a bombshell. He opened with a doomsday pronouncement that "the battle to feed humanity is over." He described the convulsions of a "dying planet." He prophesied that the new generation would witness the death of "the standards, politics and economics" of the 1960's. And he concluded with a Cassandra-like declaration that "hundreds of millions" of human beings would starve to death in the coming decade.

Paul's thunderbolts reverberated in many corridors. They helped spread the word that turned the 1970 Earth Day celebrations on the nation's campuses into an incredible, one-day exercise in environmental education. They threw a spotlight on global ecological problems. And they injected population issues into the national dialogue by asking searing questions, for the first time, about the impacts of further population increases on our resource future and on the quality of life in this country.

Ehrlich's magnetism captured a fervent army of followers on and off the nation's campuses. The same year *Bomb* appeared, Ehrlich and two of his friends formed a militant new organization, Zero Population Growth, to enter the arenas of political action. ZPG's growth was phenomenal: In eighteen months it had 250 chapters with 23,000 members and an aggressive lobbying operation under way in Washington. And at the state level, an Ehrlich disciple, a young legislator named Richard Lamm, was honing environmental arguments that would later help elect him governor of Colorado for three terms. In the 1980's Lamm supplanted his mentor as the nation's most effective spokesman for population control and immigration reform.

Ehrlich's sudden stardom evoked one of the few unedifying family squabbles of the initial phase of the environmental movement. Commoner would have done his new rival a favor if he had urged him to tone down some of his spectacular pronouncements about impending famines. However, he directed his fire at Ehrlich's core thesis (that it would be impossible for third world nations to sustain significant economic growth unless they first reduced their rates of population growth) by turning this maxim on its head and arguing that the only way poor countries could control their spiraling populations was through economic advances that would make "demographic transitions" feasible.

Ehrlich's 1968 appearance as the first TV superstar of the environmental movement was a commentary on the excitement ecology was generating at the end of the 1960's. For a time, the two disparate lives Paul Ehrlich led symbolized this volatility. There was Ehrlich the showman who was unfairly viewed by some of his peers as a scientific charlatan. This Ehrlich wrote *Bomb*, coined new words such as *ecotactics* and *ecocatastrophe*, and threw out chilling doomsday prophecies during his frequent appearances on Johnny Carson's "Tonight Show." The other Ehrlich was a superb biology teacher who had a prolific pen and an avid interest in original research. He spent part of his spare time writing incisive science books and another part executing finely honed scientific studies of the environmental impact of modern technology.

One can follow the two lives of Paul Ehrlich through the works he produced during the first seven years of his dual career. In 1969-70 he and his wife, Anne, a gifted Stanford biologist, wrote *Population/Resources/Environment*, which became the leading college textbook for environmental studies. In 1971 he collaborated on *How to Be a Survivor*, a quickie sequel to *Bomb*, which told the younger set how to "save spaceship Earth." The same year he co-edited a selection of scholarly essays, *Man and* the *Ecosphere*, and began work on another classroom text, *Introduction to Biology*, published in 1973.

Ehrlich's versatility came to a climax in 1974 with the publication of three volumes. For the aficionados of *Bomb* there was *The End of Affluence*, Ehrlich's final pop-ecology paperback, featuring fresh predictions of a billion or more deaths in "new famines," a prophecy that Japan was a "dying giant," and a "blueprint" to help readers surmount the "great dislocations" of the coming decades. Ehrlich also co-authored a book, *Ark II: Social Response to Environmental Imperatives*, that laid out a smorgasbord of ideas for social and political change. And the same year, he collaborated with some colleagues on a more scholarly tome, *The Process of Evolution*.

But at the very time Barry Commoner was plunging into politics, Paul Ehrlich apparently decided to discard his prophet's cloak and devote his main energies to science. He dabbled in sociology with two subsequent books, *The Race Bomb* and *The Golden Door*, but in the 1980's the bulk of his research has focused on studies of the strains placed on earth's ecosystems by the surging engines of Big Technology.

Ehrlich's sustained work in the vineyards of science ultimately quieted his colleagues' criticisms. When a distinguished international panel of biologists gathered in 1983 to assess the potential impacts of a nuclear war on earth's life-support systems, the conferees chose Paul Ehrlich to chair their deliberations and to draft the report that summarized their findings. This document, and Dr. Carl Sagan's companion summary of the atmospheric and climatic consequences of nuclear war, form the core of *The Cold and the Dark: The World After Nuclear War*, a science book that contains some of the most fateful research findings of this century.

To this day Paul Ehrlich's energy and acumen remain undiminished, and he is still making important contributions to ecological knowledge. Indeed, his 1986 book, *The Machinery of Nature,* has been described by one scientist as "magnificent . . . one of the best popular science books in print."

Earth Day

In 1988, it is indisputable that the environmental movement represents a historic upwelling of grass-roots democracy. The swift rise of Ralph Nader, Barry Commoner, and Paul Ehrlich —who all vaulted into positions of national leadership from inauspicious beginnings in Winsted, Brooklyn, and Maplewood—is a compelling statement about the ferment in American democracy during recent decades.

It was, however, the Earth Day celebration that affirmed the legitimacy of the environmental movement's claim to status as a powerful and permanent new force in American life. Wisconsin Senator Gaylord Nelson conceived an Earth Day as a teach-in to inform the new generation about ecological truths. With a federal grant of $125,000, three Harvard graduate students set up a network to provide information to the do-your-own-thing committees that volunteered to participate in this event. The outpouring of student interest exceeded the senator's wildest dreams.

On April 22, 1970, rallies, demonstrations, and protests were held at over 1,500 colleges and 10,000 elementary and secondary schools in fifty states. In a single day, lavish national and local television coverage created a landmark in communications history by raising the environmental consciousness of tens of millions of citizens.

This educational boostershot was a milestone for the environmental movement. It exposed so many Americans to the subject of ecology for the first time that some latecomers concluded that the environmental movement began in this country on Earth Day. But that was bad history (like asserting that the U.S. civil rights movement began at the Lincoln Memorial on the day Martin Luther King, Jr., gave his magnificent "I had a dream" speech), for it ignored the nurturing process that had brought ecological truths into the mainstream of American thought. But Earth Day was indeed a red-letter day, a coming of age for the American environmental movement.

CHAPTER XX

Confronting the Problem of Human Restraint

Man's history has thus entered a new age ... The man-made world has become not only gigantic and overwhelming, but sometimes even monstrous.

—AURELIO PECCEI, Founder
The Club of Rome

It is literally fantastic to realize the amounts of effort and money we are spending on mechanistic developments ... at the same time we neglect our irreplaceable inheritance of life from the past ages.

—CHARLES A. LINDBERGH
"Letter to Stewart Udall" (1967)

THE PROSPECT that humankind was on the threshold of a technological wonderland first emerged when prestigious scientists, heralding the Atomic Age, announced that the harnessing of "peaceful" atoms would provide electricity so cheap it wouldn't have to be metered. This pronouncement and a subsequent international Atoms for Peace conference sponsored by the United States in Geneva in 1955 altered the outlook for the whole human enterprise. These developments offered assurances that any limits to growth could be removed. They signaled an end to resource shortages that had long hindered human advancement. And they vouchsafed that scientists would synthesize cheaper, superior resources as our old resources, such as petroleum, were depleted.

The American people embraced those optimistic projections. The awe and secrecy that enveloped nuclear issues in those days meant that few ordinary citizens had either the knowledge or the temerity to ask probing questions about these new world scenarios. The ebullience whipped up by these futuristic blueprints turned faith in technology into a kind of secular religion as the "soaring sixties" began. To a large degree, it was this surge of technological optimism that persuaded our leaders that the United States could simultaneously go to the moon, feed the world's hungry, launch a program to modernize Latin America, win a war in Southeast Asia, and provide a military shield for the free world.

In the 1960's, President Kennedy's Apollo program became a TV extravaganza that dramatized the new reach of American power. In reality, NASA represented a superb, but narrow, engineering achievement; but it also became a great hype machine that used each exploit in space to construct new myths about the world-changing potentials offered by science and technology. It was these same NASA folk who polished and popularized the slogan, "Today's science fiction is tomorrow's reality," as they described experiments that would lead to such projects as mining the moon, manipulating the earth's weather from space platforms, exporting polluting industries to asteroids, mounting shuttle trips to other planets, and constructing colonies in outer space to serve as "backup stations" for earth's inhabitants.

America's euphoria over the new world our engineers were creating came to a climax with a flurry of second-coming rhetoric when our astronauts landed on the moon in July 1969. Rocketeer Wernher von Braun pontificated that space exploration was "the salvation of the human race"; President Richard M. Nixon exclaimed that the American moon landing marked "the greatest week since the creation of the earth"; and this achievement inspired the boast, "This proves that this nation can do whatever it decides to do!"

A mastery that promised a superabundance of resources—and eliminated the requirement for any restraint—was a central theme of the superoptimism generated by our "conquest" of space. Herman Kahn, the leader of a new breed of futurologists, pointedly belittled the importance of the earth and its resources by asserting that "the only real resource is the mind of man." Kahn's visions of economic utopias made him a celebrity in the business community. He offered his audiences an expansive description of the gains technology had already achieved for the human race: "There is no persuasive evidence that any meaningful limits to growth are in sight [and] . . . if any long-term limits set by 'finite earth' really exist, they can be offset by the vast extraterrestrial resources and areas that will become available soon."

A similar theme had been expounded earlier by R. Buckminister Fuller, an engineer noted for his bold ideas and innovative designs. Fuller, whose writings and lectures also reached a national audience, preached that a mechanistic millennium was at hand. In 1967 he disparaged the concerns of conservationists about resource shortages if population increases were not controlled by declaring that "Humanity's mastery of vast, inanimate, inexhaustible energy sources . . . has proven Malthus to be wrong. Comprehensive physical and economic success for humanity may now be accomplished in one quarter of a century."

The glowing optimism spawned by visionaries such as Kahn and Fuller not only permeated American thinking, but spilled over into the international sphere as well. In his call at the United Nations for a global decade of development to eliminate poverty and want, Director General U Thant echoed Fuller's thesis by stating, "It is no longer resources that limit decisions. It is the decision that makes the resources." U Thant envisioned that technology had propelled mankind into an era in which, "The truth, the central stupendous truth, about developed countries today is that they can have—in anything but the shortest run—the kind and scale of resources they decide to have."

This was the philosophical backdrop when environmentalist thinkers began questioning these verities in the mid-sixties. Their words were like flute music at a speedway, but two of the "musicians" who offered counter themes were the biologist, Garrett Hardin, and the iconoclastic economist, Kenneth Boulding.

In an elegant 1966 essay entitled "The Economics of the Coming Spaceship Earth," Boulding reminded his fellow economists of the earth's iron laws, deplored the trend toward "cowboy economics," and counseled his colleagues to question their assumption that in a linear economy ever-rising curves of consumption could continue indefinitely. Hardin attracted attention around the same time with a landmark essay on nature's limits, "The Tragedy of the Commons," which presented a classic argument for population control and against purveyors of the doctrine that science could produce technical solutions to all resource problems. Garrett Hardin was an incisive, witty writer who enjoyed his role as one of ecology's most adroit polemicists. His subsequent books would make him one of the environmental movement's most provocative thinkers.

It was, in the end, leadership provided by ordinary citizens—and the publicity it generated—that put restraints on certain kinds of technological "progress." During the early sixties a new strain of conservation activism appeared when irate citizens in northern New Jersey thwarted a plan of the political juggernaut known as the New York Port Authority to pave New Jersey's Great Swamp and convert it into a superjetport. And another angry band of preservationists led by Illinois's indefatigable Senator Paul Douglas started a struggle with some of the titans of the steel industry that, in due course, saved remnants of the Indiana Dunes as a National Lakeshore. In addition, a surge of citizen-led conservation coalitions in Massachusetts, Long Island, Texas, and the San Francisco Bay area beseeched Congress to purchase unsullied strands and headlands of Cape Cod, Fire Island, Padre Island and the Point Reyes Peninsula as new National Seashores for all Americans.

These kinds of coalitions were exerting political leverage at the local, state, and national levels in the mid-sixties. The goals of these ad hoc organizations were manifested by campaigns to stop the SST, to curb the manufacture of detergents that were suffocating lakes and rivers, and to halt the construction of inner-loop segments of the interstate highway system that would disrupt and destroy the ambiance of cities like San Francisco, Memphis, New York, and Washington, D.C.

In 1969 this grass-roots activism took on a new dimension that changed history when indigenous groups began organizing spontaneously in seven states to oppose previously untouchable projects of the Atomic Energy Commission. At the outset, these campaigns were not under the aegis of national environmental organizations. Rather the initial impetus came from local groups of scientists and citizens who began appearing at AEC hearings to ask probing questions about nuclear schemes that appeared to threaten their communities and surrounding environs.

In this country, an activist who stands and fights to protect the local environment is often called a NIMBY (not-in-my-backyard). The antinuclear NIMBYs began as environmental minutemen and women who conducted investigations of the potential environmental impacts of proposed installations or experiments, and later presented briefs against particular projects at public hearings. In three short years the isolated efforts of these NIMBYs served as a catalyst for political action in the nation's capitol. The ripples of concern they generated launched a national crusade, brought the big environmental organizations into the fray, and marshaled facts that won the interest and support of powerful members of Congress.

The revolt of the NIMBYs began with grass-roots meetings of concerned neighbors who were troubled by the radiation risks associated with nuclear projects. A Pennsylvania group, for example, organized to oppose the construction of a plutonium-fueled Liquid Metal Fast Breeder Reactor in Wyoming County. In the Midwest, nine Minnesota conservation organizations united to oppose the issuance of an operating license for a nuclear power plant at Monticello on the Mississippi River north of Minneapolis. And a coalition of six conservation organizations who were concerned about a threat to Lake Michigan's fishery intervened to fight the granting of a permit to operate a nuclear plant on Lake Michigan's shoreline.

Elsewhere across the continent in 1970 and 1971 other NIMBY cadres were also mounting campaigns. In Maryland an environmental posse called the Chesapeake Environmental Protection Association started a legal skirmish over a construction permit for a bayshore power plant at Calvert Cliffs. The farming town of Lyons became the focus of a NIMBY committee that ultimately thwarted an AEC plan to situate a national repository for millennial nuclear wastes in a Kansas salt bed. And the first stirrings of two pitched battles that would persist for over fifteen years commenced when NIMBY study groups on Long Island and in San Luis Obispo, California began raising paramount safety issues about the siting of the Shoreham and Diablo Canyon nuclear power plants.

These dogged ad hoc associations were the vanguard of the antinuclear movement in this country. They pioneered the right of citizens to know the facts about radiation risks, and they demonstrated that if citizens did their homework, they could participate effectively in the process by which decisions are made about atomic projects. And, more important, their quests for facts became a movable classroom that educated the nation about the benefits and risks of the nuclear energy option.

It was the NIMBYs who first stripped away the curtains of secrecy that had impeded full and frank discussion of nuclear issues. For over two decades, nuclear scientists had worked in an Oz-like environment where their press releases were treated as solemn facts, and it was considered impertinent for ordinary citizens to ask questions once the high priests of atomic science had proclaimed that the processes and projects they had sanctioned were "clean" or "safe." The NIMBYs' findings also demonstrated to the nation that it was both risky and unwise to presume that the Atomic Energy Commission could or would serve as chief planner and propagandist for "the friendly atom" while simultaneously protecting the unknowing public from radiation hazards no one understood.

Their efforts spanned the continent, and in three short years the uncoordinated campaigns of the various NIMBY citizen groups—in spite of the absence of formal organization beyond the local level—served as a catalyst for political action in the nation's capitol. Their successes brought the big battalions of the environmental movement into the fray, encouraged powerful congressmen such as Maine's Edmund Muskie to hold hearings on radiation pollution, and provided forums where, for the first time, distinguished "heretic" scientists (such as Dr. John Gofman) could present their research findings about radiation dangers.

More than any other campaign by environmentalists, the NIMBYs' revolt affirmed the growing political savvy of the ecological community. In what can only be described as a remarkable exhibition of grass-roots democracy, in five swift years they persuaded Congress to dismantle that once-sacrosanct symbol of Big Science, the Atomic Energy Commission. And they did more than stop ill-conceived projects: their work helped introduce an element of restraint into American thought about the future of Big Technology. The NIMBYs were pioneers who insisted that even the miracle projects of the great men of nuclear science had to undergo environmental scrutiny—and in 1972 the editor of the *Bulletin of the Atomic Scientists* described the cumulative impact of their diverse campaigns as "a crusade unprecedented in the history of technology."

When NASA officials were presenting the conquest of outer space as the paramount challenge facing the American people, unfortunately there were no antispace NIMBYs to pose questions. The space program did not present a physical threat to earth's environment; nevertheless, most environmentalists resented NASA's refrain that somehow humankind's future lay beyond the confines of earth. After the moon landing, their resentment finally boiled over when a high NASA official pontificated that "Should man fall back from his destiny . . . the confines of this planet will destroy him."

The environmentalist who responded was the scientist-poet, Loren Eiseley. "But where," Eiseley wanted to know, "did these men intend to flee? The solar system stretched bleak and cold and craterstrewn before my mind. The nearest, probably planetless star was four light-years away and many human generations away . . . No, I thought, this planet nourished man . . . Space flight is a brave venture, but upon the soaring rockets are projected all the fears and evasions of man . . . Earth will not destroy him. It is he who threatens to destroy the earth."

In 1972 support for global policies of restraint arrived from an unexpected quarter—an international study group that called itself the Club of Rome. This unique study group of scientists and businessmen was formed by Aurelio Peccei, a thoughtful Italian industrialist.

Peccei began to worry about postwar growth trends, and in 1968 he assembled seventy-five scientists and businessmen from many nations to analyze "the predicament of mankind" if existing patterns of growth persisted. The club decided to finance a study that would use computers to analyze the impacts on the earth's environment if current patterns of expansion continued for a century. This was inherently an ecological analysis, for it sought to measure all of the effects and interactions on nature's system as a result of unbridled increases of population, production, pollution, and resource extraction.

The club's report, *The Limits to Growth*, concluded that limits were on the horizon, called for "a Copernican revolution of the mind," and urged the world's leaders to deal with this reality and begin a "controlled, orderly transition from growth to global equilibrium." Peccei and his associates were casting a resounding vote for restraint by rejecting the rosy assumptions of technologists who had earlier convinced the world's leaders that all limits to growth could be "removed." The club's book was published in thirty languages, was bought by four million readers, and aroused controversies that brought the vital nature of ecological concerns into the minds of the world's leaders.

But the event that ultimately punctured the smugness of these same leaders was the impact on the global economy of the embargo imposed by a group of oil-producing nations (OPEC) twenty months later. This bold thrust for power by a group of so-called developing nations served as a rude reminder that cheap petroleum—not technology—was the resource sustaining the intrastructure of modern civilization. The United States, for example, awoke to the reality that despite all of its proud talk about technological breakthroughs, it was still a country that ran on oil. And that dependency was growing every year. In 1973 the U.S. consumed over 30 per cent of the free world's oil, was the globe's largest importer of oil, and faced the sobering fact that its reserves of crude oil were, for the first time in history, dwindling.

Washington's response to its first energy crisis revealed the mind-set that had been molded by our technological optimism.

President Nixon issued a call to action that was, in essence, a demand for a quick fix by the nation's technicians. Terming his plan "Project Independence," Nixon invoked "the spirit of Apollo and the determination of the Manhattan Project," and ordered this country's technologists to increase domestic production so the U.S. could be energy self-sufficient by 1980. There was no hint in this grandiose presidential plan that conserving petroleum by eliminating the wasteful uses of this precious commodity might be part of the answer. Indeed, when it was suggested that his country was guilty of gluttony in using over 30 per cent of the free world's oil, Mr. Nixon crisply replied that such massive consumption was a sure symbol of national strength!

Certain that conservation was necessary for the nation's economic salvation, environmentalists had contrary convictions from the outset of the embargo. In the 1970's, conventional economists clung to a dogma that economic expansion and energy use had to move in lockstep on an ever-upward curve if the country's gross national product was to grow. However, the ecologists demonstrated that this anticonservation dogma was unsound as efforts to conserve energy got under way during the next decade.

Since Americans believed that science could solve the energy problem, little thought was given to energy efficiency prior to OPEC's rude knock on our door. The two ecologists who stepped forward in the 1970's with much-needed concepts about energy thrift were an idiosyncratic British economist, E. F. Schumacher, and a twenty-nine-year-old American physicist, Amory Lovins.

Fritz Schumacher, a pacifist farm laborer in England during World War II, later became the National Coal Board's chief economist and subsequently acquired expertise in third-world development while serving as a consultant for the governments of Burma and India. His field work alerted him to the folly of the West's effort to impose its energy-intensive megasystems on the economics of countries that were at a simpler stage of development. Schumacher evolved an innovative, people-oriented concept of gradual growth geared to what he called "appropriate technology." An organizer with a humanistic touch, he readily concluded that the way to help people help themselves was to respect the limits of their cultures and offer them access to technologies that fit their immediate needs.

Schumacher summarized what he had learned through helping new nations experiment with intermediate technologies in a widely read 1973 book, *Small Is Beautiful.* As a philosopher of a "nobler economics," he argued that the "purification of human character is the essence of civilization, not the multiplication of wants." Moreover, while Dr. Schumacher gloried in the "brilliant scientific knowledge and almost infinite technological possibilities" man had achieved, he still urged his fellow beings to "respect and treat with humility that which is greater than himself, the essential unity of mankind and the great, subtle, impenetrable Web of Nature."

Although he used some of Schumacher's bricks in his edifice of ideas, Amory Lovins scanned all of the energy uses of the modern world before he began prescribing reforms for both the developed and the developing nations. What gave this young physicist exceptional vision was the holistic method he used to juxtapose the existing constructs of civilization against the earth's resources. With a versatility that bordered on genius he encompassed, and intermeshed, his own training with the disciplines of economics, mathematics, ecology, and sociology to produce masterful solutions to the world's assorted energy problems.

Amory Lovins's unfettered mind allowed him to be a sprinter, and in two years he wrote what became the winning argument against the development of the breeder-reactor by the United States, prepared an incisive global study of energy alternatives, and formulated a "soft path," less-is-more conservation strategy that revolutionized thinking about the development and use of all forms of energy. This effort culminated in a widely quoted article in *Foreign Affairs,* and a seminal book, *The Soft Energy Path,* outlining his prescriptions for authentic energy self-sufficiency for different societies.

In a sense, Amory Lovins—with his consuming conviction that conservation is the highest and most important form of national thrift—is a sophisticated throwback to Theodore and Frankin Roosevelt. And his status in 1988 as an energy advisor to industries, to foreign governments, and to the Pentagon makes him a new generation scientist whose career exemplifies the influence of ecological thought twenty-five years after the appearance of *Silent Spring.*

CHAPTER XXI

Encounter with Reaganism

The [Reagan] Administration has no commitment to the
environment and no environmental policy.
—ANNE M. BURFORD
President Reagan's first EPA Administrator (1985)

AFTER NEARLY two decades of gains under five presidents,
environmentalists were stunned in 1981 when the new Secretary of the Interior, James Watt, characterized their goals as
"extremist" and announced that the Reagan administration
intended to "reverse twenty-five years of bad resource management." But Watt's picadorish pronouncements were only the
beginning. When all of the ramifications of Reaganomics were
spelled out, it was clear that the administration believed it was
wrong to promote specific environmental goals.

Ronald Reagan's ideology about the evils of big government
impelled him to reject the national consensus about resource
stewardship that had begun with President Theodore Roosevelt.
What made Reagan the first overtly anticonservation president
of this century was the belief he shared with President Calvin
Coolidge that "the business of government is business."

Reagan centered his attention on a short list of core concepts.
His predominant precept was the conviction that if stifling intrusions of the federal government were minimized (if, in his pet
phrase, the government "got off the back" of American business) the operation of free markets would propel the nation

into a new era of affluence and power. It was this concept that encouraged the Reaganites to dismiss national stewardship as an outdated idea: planning and action to manage the nation's resources and secure environmental benefits for future generations would simply be superfluous once their free market cure-all was functioning.

Of course, environmentalism was not the only target of the administration's ideology. The dogma that has dominated the Reagan years is characterized by a conviction that the federal government—with the notable exception of the Pentagon—was a bloated bureaucracy that had to be shrunk; a tenet that many environmental laws and regulations were choking industrial growth; a thesis that dynamism animated by unregulated markets, not initiatives by governments, would ameliorate—and ultimately resolve—the nation's social and environmental ills; a dogged affirmation that tax increases are always a threat to the national welfare; a credo that the states were better situated than the federal government to manage the nation's resources and help citizens cope with social problems; and a belief that the operation of free markets would solve the nation's energy predicament.

This ideological reorientation accounts for President Reagan's zeal for deregulation and his assiduous efforts to shrivel the functions of most nonmilitary agencies. It also illuminates the logic behind the Reaganites' knee-jerk assumption that every environmental gain would burden the economy with a corresponding economic loss for business. And it explains why Reagan's first Director of the Environmental Protection Agency, Anne Burford, began her incumbency with the assumption that EPA's main mission was not to protect the environment, but to cushion the impact of "unreasonable" laws on the industries it was her duty to regulate.

The same assumption guided Interior Secretary Watt's drive to "restore resource management to the people." Watt's rapport with Ronald Reagan emboldened him to challenge the "power and self-righteousness of Big Environmentalism" by flouting existing environmental laws as he carried out his plans to increase drilling, mining, and logging on public lands. And to dramatize his disdain for environmentalists' concerns about the potential impacts of ill-planned offshore drilling, Watt ordered an unprecedented one-billion-acre sale of oil leases on the continental shelf.

Using national security needs as his cloak, Watt delegated authority to the states to police strip mining and launched an effort to inventory all mineable minerals in areas that were being considered as additions to the nation's wilderness estate. And he served notice on the American people that even popular conservation programs had to be sacrificed by announcing the phaseout of the Land and Conservation Fund, which for two decades had financed an unprecedented expansion of federal and state park, wildlife, and outdoor recreation programs.

The Reagan agenda was given momentum also by administrative changes designed to alter the "anti-industry" policies of previous administrations. In some instances environmental regulations were relaxed, and in others tacit decisions were made that laws on the books would not be enforced. In the case of the Superfund statute enacted by Congress in 1980, EPA officials spent so much time cozying up to the chemical companies that for several years nothing was done to clean up the dumps of toxic chemicals that were threatening the health of many communities. And despite the safety fiasco at Three Mile Island, the Nuclear Regulatory Commission was packed with proponents of atomic power who were so intent on accelerating the licensing of nuclear power plants—and on minimizing the implications of nuclear "accidents"—that they showed little interest in performing their duties as regulators of radiation safety.

But it was in the area of energy economics that Ronald Reagan's nonstewardship stance did the greatest harm to the nation's future. Leadership by presidents and by Congress in the 1970's made it possible for the country to pass rigorous energy conservation laws and conduct successful campaigns to reduce its dependence on imported oil. These efforts encouraged the American people and U.S. industries to make decisions about homes, autos, public transportation, and industrial plants that saved hundreds of billions of dollars and put the nation on a path of energy efficiency.

During the Reagan years the government abdicated leadership of this effort. Federal funding for energy research was drastically cut, the rules that compel Detroit to produce more efficient autos were relaxed, and the Reaganites not only sought to dismantle the energy conservation programs they had inherited, but to systematically oppose new legislation fostering energy efficiency. Reagan's decision that the open market, not federal leadership expressed through energy conservation laws and programs, should determine the nation's energy policy is a judgment that is now setting the stage for an economic disaster in the 1990's as our commitment to end energy waste slackens and our dependence on imported oil increases each day.

President Reagan's stance that the nation's decisions about resources should be made in the marketplace made it difficult for conservationists to develop a response to the new administration. The President could have provoked a meaningful national debate had he confronted Congress with legislation to repeal or revise the laws his administration disliked. He might, for example, have urged that the reach of the Wilderness Act be cut back, or that the authority of the Environmental Protection Agency be circumscribed. He could have submitted bills to "return" most of the nation's public land to the states or recommended the outright repeal of statues such as

NEPA and the Endangered Species Act which the White House viewed as either obstructive or absurd.

Such blunt proposals would have precipitated a meaningful national debate, but the administration's tactical decision to avoid such showdowns meant that many vital issues would be contested on two battlegrounds. The first was the budgetary arena where the administration attempted, with considerable success, to achieve a de facto repeal of "bad" laws by using the President's budget to strangle resource programs it opposed. The second was the courts, where lawyers representing the EDF, the NRDC, and the Sierra Club's Legal Defense Fund filed lawsuits to force officials of recalcitrant federal agencies to obey and enforce existing laws. After 1981, some of the best talent of the environmental movement was used to persuade federal judges to compel Secretary Watt and his successors, and other Reagan appointees, to faithfully execute important laws.

The environmental movement's overall response to the Reagan revolution demonstrated resilience and staying power. The alarms sounded by Reaganites swelled memberships of the national organizations from four million in 1981 to roughly seven million in 1988. Adversity encouraged groups to form new alliances at the national level and fostered the formation of aggressive local coalitions to counter anticonservation initiatives emanating from Washington.

Moreover, while the Reaganites were dragging their feet on the protection of endangered species, the "quiet man" of the environmental movement, the Nature Conservancy, partially filled the gap by becoming the nation's prime acquirer and protector of the habitats of endangered plants and animals. And problem-solving initiatives of versatile organizations such as EDF and NRDC enabled them to outperform the federal government in helping embattled cities, states, and industries develop innovative solutions to regional water, energy, and other resource problems. It is ironic that an administration which disavowed the concept that each generation owes a moral duty to future generations actually galvanized Americans to higher levels of commitment and action to save earth's ecosystems.

It is clear that on environmental issues Ronald Reagan rowed against the American mainstream for eight years. The "great communicator" was unable to persuade Congress to repeal a single important law he disliked. He is the only president this century who served his term without proposing any major initiative to further the cause of conservation. And all available evidence suggests that he did little to negatively influence American attitudes on ecological matters. Indeed, a 1987 public opinion poll revealed that environmental quality "has again become a consensual issue . . . as it was in the early 1970's."

I am convinced historians will one day indict the Reagan administration for its lack of vision concerning resources and its abdication of the traditional U.S. role of leadership in global environmental matters. Moreover, history will confirm that Ronald Reagan's legacy created a massive fiscal debt restricting the options of his successors and of the American people for positive action on behalf of their air, water, and land.

CHAPTER XXII

Ecology and the Future

We have not inherited the earth from our fathers, we are borrowing it from our children.
    —LESTER BROWN
        Building a Sustainable Society (1981)

IN THE quarter-century since *Silent Spring* appeared, a sophisticated, assertive environmental movement has emerged and produced profound changes in our national life:

- Measuring and mitigating ecological impacts of governmental and industrial activities is now an integral part of our public policy-making.
- Scientists have developed and standardized techniques to monitor the inroads of pollution and the subtle functioning of ecosystems.
- Innovative lawyers and jurists have brought actions of industries and governments under the rule of law.
- New knowledge about disease-causing pollutants has broadened the conservation effort into a campaign for a health-giving environment.
- Park and wildlife programs championed by conservationists have more than doubled the size of our estate of national lands.
- Congress has put a protective legal mantle around the nation's wildlands, and has adopted policies to preserve the habitats needed by endangered species.

- Scientists studying the ecological consequences of overpopulation have heightened awareness of the links among population pressures, the maintenance of natural diversity, and the quality of life for all of earth's inhabitants.
- Campaigns for energy efficiency, though sporadic, have produced impressive savings by altering modes of energy consumption in the automobile, electric power, and household sectors of our economy.
- And, finally, the diffusion of ecological knowledge has enlivened our democracy by encouraging ordinary citizens to participate in decision-making previously considered the province of "experts."

There are many ways to measure the influence of ecological thought on our national life. The realism it fostered has dampened optimism about technological "miracles" and the advent of an era of superabundant resources. Its imprint can be seen in the declaration by an organ of the electric power industry that "the environmental ethic is now firmly established in our nation." Its political clout can be gauged by the near-unanimous vote by which Congress rebuffed a presidential veto of a clean-water bill in 1987. And its social importance can be assessed by the attention accorded environmental issues by the media.

As we look ahead, the twin themes of conservation and sustainable growth will be in the forefront as the nation confronts the environmental challenges it will face in the final decade of this century. I believe, for example, that our response on one crucial issue—the rejuvenation of the Land and Water Conservation Fund—will test whether the nation wants to further expand its green legacy of park and wildlife lands. Farsighted leaders of the Nature Conservancy have warned, for example, that a last-ditch fight to save habitats for this country's remaining endangered species of plants and animals will be won or lost in the next ten years. A once-in-a-century opportunity will be lost unless Congress endows a permanent fund that will generate at least a billion dollars each year to purchase lands needed to sustain the spaciousness and ecological diversity of the American environment.

It is also significant that civic leaders of urban areas blighted by air pollution are now belatedly recognizing that revolutionary, city-changing strategies are needed to cleanse their air. The beginning of wisdom can be seen in the growing realization that the problem cannot be alleviated simply by tinkering with the tail pipes of automotive vehicles. City, state, and federal governments must join together in the 1990's and pursue total-environment strategies involving cleaner combustion engines, cleaner-burning fuels, huge investments in public transportation, and land use controls that curb urban sprawl.

Moreover, it is unreasonable to expect that cities running out of landfills for the resources we call wastes will be able to devise satisfactory solutions unless industries and state and local governments develop new recycling strategies. My generation has spoken glibly about creating a "recycling society." But now, as a garbage glut threatens to overwhelm many cities, the country has an opportunity, through environmental leadership and creative engineering, to put new systems in place that will teach Americans how to replace the excrescences of their throwaway culture with habits and values that inculcate thrift and reclamation.

The handling and disposal of toxic and radioactive residues produced by our chemical and atomic industries is another challenge facing the next generation. These wastes, some of which will present hazards to humanity for thousands of years, cast long shadows over future generations. If environmentally acceptable disposal solutions are to be devised, industrial scientists and engineers must develop techniques to minimize the output of future wastes at the same time they master the art of safely storing existing wastes in secure, impermeable niches of nature.

Perhaps one of the most fateful domestic environmental and social problems in the next decade, however, is the circumstance that this country's automobile culture, as it exists today, is not sustainable. We have invested trillions of dollars erecting an infrastructure of homes, shopping centers, workplaces, and highway systems wholly dependent on the availability of low-cost motor fuels. Yet, the prospect is increasing each week that the inexorable decline of domestic petroleum production will combine with the enormous debts our nation is contracting to cripple our ability to pay for oil imports needed to maintain our one-person one-car civilization.

Seven of America's thirteen largest oil fields are 80 per cent depleted, and we face the prospect that by 1995 nearly two-thirds of the crude oil we need will be imported. If present patterns of consumption continue, economists are projecting an annual bill for imported oil of about $150 billion by 1995. Simple arithmetic tells us that our economy cannot sustain such a burden. The social and monetary reverberations of such a permanent oil shortfall would be severe, for it would be disastrous if large segments of our infrastructure were suddenly rendered obsolete. Such a development would also shatter the freewheeling lifestyle that has seen Americans (who own 40 per cent of the world's autos) travel almost as many auto-miles each year as the other 95 per cent of earth's inhabitants.

Such an outcome could be avoided, of course, if this country were hard at work developing alternate fuels, or if drastic action to curb unnecessary travel, such as a stiff, European-style gasoline tax, were contemplated. In the absence of such action, America is drifting toward an economic Niagara, and each year brings us closer to the brink. But our inaction to date—and the unwillingness of our political leaders to confront this issue—

offers little hope that we will avoid traumatic social and environmental dislocations.

The wholesale renovation of our cities and their transportation systems would be both daunting and dispiriting were it
not for the fact that such a restructuring would confer enormous economic and environmental benefits. Our extravagant
romance with the automobile has always been a mixed bless-
ing. A transformation that would, at once, make our cities more
intimate and compact, cleanse their air, strengthen the values
that bind neighborhoods and communities together—and give
priority to convenient public transportation—could generate
wide-ranging changes that would revitalize our cities.

Once events force them to come to terms with the energy
imperatives of the 1990's, and they realize that mobility must
be sacrificed as our communities and transportation systems
are reshaped, the citizens of this country will face a great challenge. Will a new breed of politician emerge who can use energy
security and national efficiency as grand themes of a latter-
day New Deal? Will the American people respond resiliently
when they realize that their freewheeling days are over and they
must redesign their lives, their cities, and their lifestyles? These
are environmental and resource issues that have the potential
to dominate our domestic agendas in the twenty-first century.

With U.S. organizations providing much of the dynamism,
the international environmental movement has exerted a growing influence on global decision-making in the 1980's. This
should be applauded, for it reflects a scientific consensus that
common efforts are essential to protect the natural systems that
sustain all life on this planet. Issues such as ozone depletion
and the climate-changing greenhouse effect, which concerned
only a few specialists a decade ago, have been propelled onto
the agendas of organizations like the United Nations and the
World Bank.

This attests to both the power of the environmental movement and the swift spread of knowledge about ecological truth. The eminent geographer, Gilbert F. White, commented on these "truly radical shifts in our perceptions of the world environment" and expressed wonderment at the speed with which "a universal environmental ethic" has spread around the globe and gained disciples among both laymen and leaders in the contrasting cultures of the world. Ecology, in one generation, has been transformed from a cause of an elite to a central concern of humankind. The new password is, "Think globally, act locally."

Recent scientific investigations into the impacts of world-wide industrial activities on the ecosphere have aroused profound concerns about the preservation of the planet's natural systems. These findings have galvanized action to curb acid deposition, to protect earth's ozone shield, to save the world's remaining rain forests, to halt deforestation and desertification, to check the rampant extermination of endangered species, to help nations bring their populations into a sustainable balance with their resources, and to devise ways to cope with an accumulation of greenhouse gases that could change rainfall patterns, melt ice caps, and inundate some of the globe's great cities.

Humankind cannot ignore this intimidating agenda. The reason is simple: no single nation can effectively resolve these problems. Obviously, the scale of actions needed will require unprecedented international cooperation and goodwill during the next century. But is such cooperation possible? Two developments in 1987 suggest that it might be.

In a remarkable display of cooperation, twenty-four industrialized nations met in Montreal to approve a treaty to restrict production of the chlorofluorocarbons (CFCs) that have been dissolving earth's shield of stratospheric ozone. The most hopeful thing about this accord was that scientists affiliated with the United Nations Environment Program and environmental organizations in the U.S. and Europe provided the leadership that brought it to fruition.

Initially, politicians both here and abroad sought to protect the investments of their chemical companies by ridiculing this initiative. Interior Secretary Donald Hodel, for example, urged U.S. citizens to ignore scare warnings about epidemics of skin cancer. Hodel advocated inaction, and proposed that everyone "wear dark glasses and broad-brimmed hats." However, a few months after the Montreal meeting, creative chemists had the last word when the American Telephone and Telegraph Company announced that it had developed a cost-competitive substitute for CFCs. And DuPont, the leading CFC manufacturer, has announced it will accelerate the phaseout of this chemical.

The other positive note was provided by a team of Natural Resources Defense Council scientists who demonstrated that advances in the art of seismic verification now make it feasible for the United States and the USSR to verify from great distances agreements banning the testing of all nuclear weapons. This was a precedent-setting example of a conservation organization taking the initiative to mediate a longstanding dispute between the superpowers. With this one stroke, NRDC furthered the cause of peace and enhanced the global prestige of the environmental movement.

The recognition that many environmental problems have a global overlay has made Greenpeace International the fastest growing new environmental organization over the past decade. By confronting despoilers on the high seas and in remote geographic regions of the world, Greenpeace has emerged as an aggressive new force for ecological reform. The bold leadership of Greenpeace's young founders—David McTaggart, Robert Hunter, and Patrick Morre—offers further evidence that the environmental movement is developing the vigor and dynamism it will need in the coming decades to tackle the globe's overarching problems.

Ecological insights constantly remind us that the resources humankind needs for the long haul can only be husbanded if we nurture an ever-widening concept of land stewardship. We must inculcate in our children a feeling of belonging to a community that is larger than any nation, more spacious than any culture. As the poet Archibald MacLeish points out, the current metaphor of our existence is that we are "riders on the earth" together. The pact with nature struggling to be born requires new relationships among peoples and nations based on the most intrinsic values of all—sharing, caring, and cooperation.

During my adult years, the march of history has conferred power on human beings to modify or impair the natural processes that renew and sustain life on earth. Now even climates can be impaired, and access to the very sun rays that make this planet the one green jewel of our solar system can be obstructed by human action. The fateful challenge facing tomorrow's environmentalists is to reach across the artificial barriers erected by nation states, languages, and cultures and become earthkeepers who steadfastly use their talents to nourish all causes that promote life on this planet. That, for the next generation, is the ultimate message of ecology.

TIMETABLE OF THE
AGE OF ECOLOGY (1962-1988)

Year	Significant Events	Significant Publications
1962		Rachel Carson. *Silent Spring*
1963	US-USSR Test Ban Treaty signed	
1964	Wilderness Act passed	Leo Marx. *The Machine in the Garden: Technology and the Pastoral Ideal in America*
	Land and Water Conservation Fund established	
1965	Storm King lawsuit won by the Sierra Club	
1966	Clean water legislation passed	Kenneth Boulding. *The Economics of Spaceship Earth*
	First law enacted to protect endangered species	
1967	Environmental Defense Fund founded	Roderick Nash. *Wilderness and the American Mind*
	Air pollution legislation passed	Garrett Hardin. *The Tragedy of the Commons*
1968	Grand Canyon dams stopped by conservation coalition	Paul Ehrlich. *The Population Bomb*
	Wild and Scenic Rivers Act passed	William H. Whyte. *The Last Landscape*

1969	Santa Barbara oil spill	Ian McHarg. *Design With Nature*

1969   Santa Barbara oil spill   Ian McHarg. *Design With Nature*

Friends of the Earth founded by David Brower National Environmental Policy Act (NEPA) passed

Nuclear power opposed by NIMBYs

1970   Earth Day observed on the nation's campuses

Trans-Alaska hot oil pipeline stopped by first NEPA lawsuit

1971   SST stopped by Congress

Environmental Protection Agency established

1972   Comprehensive law passed to control water pollution   Paul Brooks. *The House of Life*

Global issues explored by UN Stockholm Conference   Barry Commoner. *The Closing Circle*

1973   OPEC energy crunch   E. F. Schumacher. *Small Is Beautiful*

Endangered Species Act passed   Joseph Sax. *Defending the Environment*

1974   Atomic Energy Commission dismantled by Congress

1976   Land Management Policy Act passed   Amory Lovins. *The Soft Energy Path*

1977   Wendell Berry. *The Unsettling of America*

1978	Carter Energy Conservation Program initiated	
1979	Three Mile Island nuclear "excursion" Second OPEC oil shock	Lester R. Brown. *The Twenty-Ninth Day* Robert Cahn. *Footprints on the Planet*
	Carter Synfuels Program enacted	
1980	Law passed to protect 100 million acres of Alaska's public domain	*Global 2,000 Report*
1981		Kenneth Boulding. *Ecodynamics* Stephen Fox. *John Muir and His Legacy*
1984	2,000 killed by leak at US pesticide plant in Bhopal, India	*The Cold and the Dark,* by a panel of scientists.
	Carter Synfuels Program abandoned	
1985	Clinch River Breeder Reactor construction halted by Congress	
1986	Chernobyl meltdown in USSR	
1987	Treaty to protect the ozone shield signed by industrial nations	

ACKNOWLEDGMENTS (1964)

IN a very real sense the members of my Secretariat and personal staff, and the resource specialists in my Department, were collaborators in the making of this book. Any Secretary of the Interior, if he relishes his job, participates in a revolving seminar on resources, and his best teachers are the broad-gauged conservation thinkers in the Department.

Invidious omissions would surely occur if I were to set out to acknowledge specific debts owed to each of my colleagues. Consequently, I can only acknowledge their imprint and express thanks for it.

The extraordinary contributions of some individuals must be recognized, however. Foremost is my friend Wallace Stegner, the historian-novelist-teacher, who joined my staff for a few weeks in the fall of 1961 and who suggested the original outline for this book; others deserving special credit and thanks are Harold Gilliam, Donald Moser, and Sharon Francis. I also owe a debt to four friends who scrutinized particular chapters and provided much needed guidance: they are the late Oliver LaFarge, Ernest Lefever, William H. Whyte, and Joseph L. Fisher. Most of the credit for the illustrations goes to another friend, Alvin M. Josephy, Jr. However, the final judgments that appear in the text were my own, and all errors of statement and interpretation must be laid to my account.

Our research was nomadic, and a complete list of the books, periodicals, and other works read or consulted would be largely useless to both critic and student. Some original sources were used, but this is primarily a work of synthesis and interpretation, the transference of old wine into new historical bottles. The object has been to tell, between two covers, the land story of the American earth and the changing land attitudes of those who have used it and lived on it. In lieu of a bibliography, I have chosen to direct the reader only to some of the main works relied on. Here, then, are brief notes to each of the chapters.

Chapter One: Some anthropologists may quarrel with certain of my generalizations. The American Indians await their Francis Parkman, an anthropologist-historian who can catch the hues and splendors of their varied cultures. Of the works available, Oliver LaFarge's writings contain some of the most sensitive interpretations of Indian history. The *Book on Indians* published last year by American Heritage, under the editorship of Alvin M. Josephy, Jr., is a perceptive and superbly illustrated volume for the general reader.

Chapter Two: Jefferson's *Notes on Virginia* and other writings that reveal his interest in the land deserve close attention. Most of the standard American history texts are seriously deficient in their accounts of our land history. Of those I consulted, *The Growth of the American Republic* by Samuel E. Morison and Henry Steele Commager is easily the best. One regrets that the late Bernard De Voto did not have time to focus his fierce gaze and trenchant pen on the land stewardship of colonial times.

Chapter Three: The books of De Voto's trilogy on the winning of the West (*The Year of Decision, Across the Wide Missouri, The Course of Empire*) contain much superlative writing. The astonishing saga of Jedediah Smith, told in a fine biography by Dale L. Morgan, *Jedediah Smith and the Opening of the West,* is a major contribution toward an understanding of the American frontiersmen. The late Walter Prescott Webb's *The Great Plains* is a definitive, colorful account of the settlement of one region.

Chapter Four: The best source for this chapter is, of course, the writings of the protagonists themselves. For my part, the best of the Thoreau biographies is that by Joseph Wood Krutch.

Chapter Five: The story of the raid on resources must be pieced together from hundreds of books and articles. This is also an area neglected by writers of American history.

Chapter Six: David Lowenthal's biography, *George Perkins Marsh, Versatile Vermonter,* gives belated and sensitive recognition to the intellectual achievements of this remarkable American.

Chapter Seven: Wallace Stegner's Powell biography, *Beyond the Hundredth Meridian,* has such insight and immediacy that it might have been written by a lieutenant who served in a Powell survey party. The reports and papers that both Schurz and Powell filed while they held office at Interior contain the essential gospel they espoused.

Chapter Eight: Pinchot's autobiography *Breaking New Ground* has revealing passages, but the story of the history of our forestlands is best recounted in *Forest and Range Policy* by Samuel Trask Dana and the more recent *Forests and Men* by William B. Greeley, one of Pinchot's foresters.

Chapter Nine: Muir's own writings reveal the man more accurately than the later books and articles of those who love him too well to give us the real person. Nearly all of Stephen Mather's papers were burned or lost. It is regrettable that we know so little about him and the pioneer period of the parklands.

Chapter Ten: Easily the best biography of T. R. is *Power and Responsibility: The Life and Times of Theodore Roosevelt* by William Henry Harbaugh. The two-volume work, *Franklin D. Roosevelt and Conservation 1911–1945*, put out by the Roosevelt Library is a storehouse of original source materials. Few of the New Deal historians have grasped the full significance of FDR's conservation commitments and achievements. An exception is Arthur Schlesinger, Jr., who had the good fortune, at an impressionable age, to be converted and baptized by De Voto himself. The section, "The Battle for Public Development," in Schlesinger's book *The Coming of the New Deal* is first rate.

Chapter Eleven: A recent book which discusses many of the persons and events encompassed by this chapter is James B. Trefethen's *Crusade for Wildlife.*

Chapter Twelve: Lewis Mumford is, of course, the nonpareil. The Mumford classics are *The Culture of Cities* and *The City in History,* but all of his writings have a conservation-of-living-values bias that make them both inspiring and indispensable. It is a mark against his countrymen that during his lifetime he has not been given an opportunity, such as Olmsted had, to put some of his ideas into practice. City conservationists can take heart from the many provocative books which have recently been published concerning the design and planning of cities.

Chapter Thirteen: The seven reports turned out at the request of President Kennedy earlier this year by the Committee on Natural Resources of the National Academy of Sciences put the current picture in sharp focus. The recent book, *Resources in America's Future,* by Joseph L. Fisher and his associates of Resources for the Future is also very illuminating.

Chapter Fourteen: If asked to select a single volume which contains a noble elegy for the American earth and a plea for a new land ethic, most of us at Interior would vote for Aldo Leopold's *A Sand County Almanac.* Joseph Wood Krutch, now an Arizona fugitive from megalopolis, and his sometime yoke-fellow drama critic, Brooks Atkinson, have also, in the best of their nature writing, been skirting the shores of Walden Pond.

ACKNOWLEDGMENTS AND
CHAPTER NOTES (1988)

An AUTHOR who tries to be pithy owes his readers some clarifications and explanations.

My friends who are dedicated urbanists may wonder whether my interest in cities as environments has flagged, and I want to tender reassurances. I have not included a new chapter covering the urban scene because in my book, *1976: Agenda for Tomorrow,* I stated my plans and hopes for improving the quality of life in America's cities. To quote Robert Frost, "They would not find me changed from him they knew/Only more sure of all I thought was true."

To other friends who may feel that I have neglected the many contributions of Lady Bird Johnson, my riposte is that the University Press of Kansas has just published *Lady Bird Johnson and the Environment.* This sensitive book-length profile by Dr. Lewis L. Gould tells the full story of Mrs. Johnson's efforts in the 1960's. This gallant First Lady did far more than plant flowers and talk about the importance of natural beauty. She raised a banner of environmental quality that enlarged the outlook of the American people.

Finally, I hope that my admittedly subjective Timetable of the Age of Ecology will encourage my readers both to argue with my judgments and to be keen observers as further events unfold.

I have checked many of my impressions against what I regard as the two best recent works by historians that cover some of the same ground traced in this volume. I refer to Stephen Fox's *John Muir and His Legacy: The American Conservation Movement* (1981) and to *Beauty, Health and Permanence: Environmental Politics in the U.S. 1955–1985* by Samuel P. Hays (1987).

In lieu of a cumbersome bibliography, in the notes appended below I have elected to mention the principal works I relied on in preparing this update.

Chapter Fifteen: The definitive book thus far about Rachel Carson's life and work is *The House of Life* (1972), by her friend and editor, Paul Brooks. Another volume that illuminates the storm of controversy ignited by Carson's final book is Frank Graham's *Since Silent Spring* (1970).

Chapter Sixteen: Because of his sudden passing, Bernard De Voto did not receive the credit he deserved for the tocsin that aroused conservationists from their postwar torpor. Those who want to appreciate De Voto's contributions should read Wallace Stegner's splendid memoir of his friend's life, *The Uneasy Chair* (1974). John McPhee's profile of Dave Brower in *Encounters With the Archdruid* is notable for its many insights into the personality and character of the Sierra Club's remarkable leader. In her trenchant, *The Fight to Save the Redwoods,* Susan R. Schrepfer has given us a series of snapshots that reveal vital interactions among the leaders of the Sierra Club during its turbulent years under Brower's leadership.

Chapter Seventeen: Bob Marshall, one of A. E. Housman's "athletes who died young," has finally received the recognition he richly deserves in a fine 1987 biography, *A Wilderness Original: The Life of Bob Marshall,* by James M. Glover. It gave me great satisfaction to have an opportunity to pay a tribute to Howard Zahniser, and I owe special thanks to Paul Oehser, Zahnie's bosom friend, for sharing materials I needed to fashion a victory laurel for his brow.

Chapter Eighteen: Professor Joseph Sax of the University of Michigan law school was the first, and perhaps the best, chronicler of the golden age of environmental law. Because he has so many vivid memories of EDF's great days, I hope that one day Charles Wurster will write a lively history of the founding years of the Environmental Defense Fund.

Chapter Nineteen: The Ralph Nader biography I found most useful was Charles McCarry's *Citizen Nader.* In his work mentioned heretofore, Stephen Fox has a vivid chapter, "The Last Endangered Species," that evaluates the contributions Nader, Commoner, and Ehrlich made during the first phase of the environmental crusade.

Chapter Twenty: In my account of the revolt of the NIMBYs, I used material from Richard Lewis's *The Rebellion Against Nuclear Power* (1972). A pungent, powerfully evocative work that delves into the roots of conservation activism—and relates the eighty-year fight to save the Indiana Dunes to a region's search for community—is J. Ronald Engel's *Sacred Sands.* Dr. Engel is a theologian who sees a struggle to save public land as a classic expression of "civil religion" in this country. Those who want to know more about Italy's latter-day Renaissance man, Aurelio Peccei, will read his 1978 book, *The Human Quality.*

I wish to express my thanks to these friends who reviewed and criticized parts of my manuscript: Paul Brooks, Michael Cohen, Henry Caulfield, Spenser Havlick, Richard Lamm, Tom Turner, Charles Wurster, and my nephew, James R. Udall. My principal debt is to my personal editor, Lee Udall. Her exquisite understanding of sentence sounds and the elements of literary style have left her mark on every page of my update.

Finally, I encourage those who want to keep abreast of the ongoing saga of ecology to maintain a membership in one of the national environmental organizations and get involved in the struggle for global policies that will preserve the ecosystems of the earth.

Index

aborigines, America's, 40
arboretum, Madison (Wis.), 169
Acadia National Park, 150, 151
action
  beginning of, 83-96
  individual, 147-158
  Reclamation, of 1902, 96, 129, 130
  Taylor Grazing, 145
Adams, Henry, (quoted) 44
Adams, John, 230
Adams, John Quincy, 73-74, 130, (quoted) 74
Adirondack Forest Preserve (N.Y.), 213
advertising
  industry, 209
  newspaper, 211
Africa, east, resource insights shared with,
  186
Agassiz, Louis, 51
agriculture
  increases in food output, 198
agribusiness industry, 195, 226
Agriculture, Department of
  forest reserves transferred to jurisdiction
    of, 103
  Pinchot, Chief Forester of, 102
agriculture, Jefferson's opinion of, 19
Airport, Dulles International (Wash., D.C.),
  170
Alaska, 3
  opportunity in, 182
  sale of, to United States, 62
Alaska Commercial Company, 63
Alaska Purchase, 63
Albany (N.Y.), 56
Albright, Horace, 151
Aldicarb, 198
Algonquians, 4
Allagash River (Me.), 182
Alleghenies (Pa.)
  an early settlement boundary, 18
  Indians in, 8
Allotment Act (1887), 11
American Association for the Advancement of
  Science, 94, 100, 228
American Civic Association, 123
American Forestry Association (AFA), 87, 100,
  148
American Forestry Congress, 103
American Forests, 154
American Fur Company, 45, 62
American heritage, xi, 22
American Planning and Civic Association, 157
American Red Cross, forerunner of the, 162
American Telephone and Telegraph
  Company, 269
Anaconda Mine, 98
"Anderson Bill," 219
Anderson, Clinton P., 218, 219, 221, (quoted)
  181
Antarctic, a scientific preserve, 187
Antiquities Act, 132
antinuclear movement, 250
"Anti-Uglies," British, 157

Apaches, 10
Appalachians
  oil-mining in the, 59
  William Bartram's exploration of the, 41
Arapooish, Chief, 6, (quoted) 5
arid lands
  conservation plan for settlement of, 89-91
  desalted water for, 187
  Powell policy for, 130
arid regions, Powell report on, 88, 89-95, 96
Arikara massacre, 34
Arizona
  canals on desert of, 4
  Grand Canyon of, 118, 132
Ark II: Social Response to Environmental
    Imperatives (Paul Ehrlich), 242
Arkansas, buffalo in, 64
Arkansas River, 142
Arnold, Matthew, 75
Ashley, William H., 32, 34
Aspinall, Wayne, 217, 219, 220, 221
Astor, John Jacob, 36, 62
atomic energy, 174
Atomic Age, 245
Atomic Energy Commission (AEC), 223, 236,
  237, 249, 251
Atoms for Peace conference, 245
Audubon, John James, 40, 42-43, 47, 53
Audubon Societies, wild birds protected by,
  64
Audubon Society, 43, 155
  National, 148
autobiography, Filson-Boone, 28, 29
automobile industry, 206, 233
autos, 161, 170, 175, 235, 265, 266
Ax Age American, 55

Ballinger, Richard A., 107
Bangor (Me.), onetime lumber capital of the
  world, 56
Bartram, John, 19, 40, 41
Bartram, William, 40-41, 43, 46, 47, 48, 52,
  (quoted) 41
badlands, homemade, 60
Baxter, Percival P., 152-153
Baxter State Park, 152
Bear Lake, Jedediah Smith's summer rendez-
  vous, 34
Bear Mountain (N.Y.), 153
beaver, 3, 32, 33, 34
  mountain men's raids on, 36, 37, 54, 62
  sale value of, 35
Benét, Stephen Vincent, (quoted) 25
Bennett, Hugh Hammond, 144
Bent brothers, 33
Bering Sea, seal hunting in the, 62-63
Beverly, Robert, (quoted) 15
Big Crow, Chief, 45
big game
  organization against slaughter of, 149
  refuges for, 128, 149
Big Raiders, 127
Big Raids on resources, 54-68, 98, 136, 144

283